# The Ecological and Economic Contexts of Paliyan Development

# The Ecological and Economic Contexts of Paliyan Development

**Bibhu K. Mohanty**

GYAN

Published, 2020

*Published by*

Gyan Publishing House
5, Ansari Road
Daryaganj, New Delhi-110002
Phone: 011-47034999, 9811692060
www.gyanbooks.com
E-mail: books@gyanbooks.com

*Printed at:* G. Print Process, Delhi.

**Cataloging in Publication Data—DK**
Courtesy: D.K. Agencies (P) Ltd. <docinfo@dkagencies.com>

**Mohanty, Bibhu K.,** 1967- **author.**
The ecological and economic contexts of Paliyan development / Bibhu K. Mohanty.
pages cm
Includes bibliographical references.
ISBN 9788121218153

1. Paliyan (Indic people)—Economic conditions. 2. Human ecology—India. 3. Sustainable development—India. I. Title.

LCC DS432.P218M64 2019 | DDC 305.800954 23

# Contents

# Preface

This book titled "The Ecological And Economic Contexts of Paliyan Development" is an anthropological study of a food gathering tribe living in the rain shadow area of Western Ghats of Tamilnadu. In the present study the case of Paliyan tribe is taken up to assess the impact of changing ecology and economy and the extent of developmental process initiated by both Government and Non-Governmental agencies along with the degree of success or failure in their efforts.

Before undertaking the research, I tried to utilize secondary sources such as reports contained in gazetteers, census reports and ethnographic accounts published in the form of monographs, scientific and professional journals. Records available in various government offices and documents maintained by voluntary organizations, archives, missionary organizations are also utilized as secondary sources of data.

Next, the focus of my attention was on the distribution of Paliyans. For this I took up reconnoitre visits and found the Paliyans to be distributed in following three areas-

1. (a) Palani hills and Sirumalai hills of Western ghats located in Dindigal district. The highest concentration of Paliyans is in this area of Palani hills.
   (b) Varshanad valley and foot hills of Western ghats in the Madurai District.
2. The Srivilliputur Taluk and Watrap Talluk of Kamarajar district which contains scattered hamlets of the Paliyans at the foot of the hills such as Shenbagathope of Srivilliputur area, Aiyanar Koil of Rajapalayam area are the next important concentration of the Paliyans.
3. The Tirunelvelli district containing Sankaran Koil and other acculturised hamlets of the sparsely populated Paliyan hamlets is the next area of dispersed Paliyan area of distribution.

But the third area of Paliyan population is so sparse and since majority of them have been converted to Christianity I dropped the third area of distribution of the Paliyans for my field work. Their number is insignificant and they have lost their traditional way of life Out of the two above mentioned Paliyan areas, one village from each area has been selected for intensive study. While selecting the two villages, criteria such as communication facilities, exposure to outside influences etc. have been very much emphasized these factors contribute substantially towards accelerated developmental process. Two villages from two areas helped in making the comparative analysis. Shenbagathope village was selected from Srivilliputur Taluk and Palamalai village was selected from Kodaikanal area of Palani hills.

The present study tries to adopt exploratory research design since detailed information are lacking as regards its history and factual information about various aspects of its culture and social structure. The few data that are available at present do not give a true picture. So, painstaking field work was undertaken to bring out this volume. The survey and case study method are used as major tools of research. Traditional anthropological techniques such as genealogical method, structured and unstructured interview, participant and non-participant observations have been amply used for the collection of data Case study method has been utilized very often when adequate factual data could not be possible.

Most of the Paliyans are very suspicious of the outsiders and therefore did not cooperate with me at the initial stage. Language was also the main obstacle in carrying out the field work. Since Tamil was not my mother tongue I had to take a long time to master the language so as to collect reliable data directly from the informants. Gaining mastery over the language helped in gaining confidence of the informants as well as the establishment of rapport for the collection of reliable data.

There were some problems in collecting accurate economic data. Majority of informants hesitated to give accurate data on their loans and other expenditure incurred during consumption of alcoholic drinks and narcotics etc. Especially those who depended on food gathering economy could not give any specific data regarding income and expenditure. Anthropological method of collecting data through observation, though very much fruitfull, it was a time consuming affair. Information on day to day income and expenditure was relatively more forthcoming in Palamalai (of Palani hills) than in the Shenbagathope of Srivilliputhur area. Never-the-less efforts were taken to get more or less accurate informations by interviewing petty shop keepers, neighbouring castes, and outsiders who employ the Paliyans as their labourers. Cross checking those data with the statement of the Paliyans, some realistic picture could emerge about the present economic situation of the Paliyans.

Details of developmental activities and financial expenditure for each project and the degree of success of such projects could not be assessed as the concerned officials did not fully cooperate with the researcher. They hesitated to give details of official information regarding development to avoid contradictions and exposing their own lacunas.

The book is spread over six chapters apart from the introduction. In the introduction the author introduces the subject to the reader by narrating the problem of study, objectives, theoretical perspective, issues from survey of literature, anthropological approach of tribal development, the role of anthropology in developmental strategy, hypotheses and methodological perspective etc.

In the first chapter an effort has been made to present a bird's eye view of traditional culture of the Paliyans during colonial period. Since Paliyans were a preliterate community no document is expected from them before the colonial period. During the colonial period brief or fairly long accounts of the Paliyans have been recorded by the administrators, missionaries, census and gazetteers compilers and ethnographers. Based on such documents a total or holistic picture of the Paliyans culture is attempted. The Paliyan culture is considered primitive, mainly because of (1) their isolated habitat in hills and forested areas, the exotic nature of their culture and (3) the negative and scornful attitude of a few authors and out siders towards them. Finally, an effort has been made to explore the history of the paliyan culture with the help of ethno-historic method. The acculturation process during the precolonial and colonial period are hinted briefly.

In the second chapter, the ecology of the Paliyan habitat is being presented. Location of their settlement, attitude and proximity to the forest, hill streams and other facilities are described. The rainfall, the flora, fauna, the exploitation of the surrounding area, the level of technology of the Paliyans and the social organization of their economic activities has been discussed here. The social structure of the Paliyans and its relation with ecology has been discussed at the end of the chapter. Finally, how the ecology influences their economic activities, health care system through herbal medicine, religious and magical practices has been elaborately analysed in this chapter.

The third chapter deals with the economic life of the Paliyans. Here the main traditional occupations are discussed. The main focus is on the two villages and how the economic life differs between these two villages. Out of these two, one belongs to Palani Hills of Dindigul district where new plantations have brought about change in the economic life of the Paliyans. Here most of the Paliyans have adopted Christianity and settled life. On the contrary, the village belonging to Kamarajar District still follows the traditional economic life due to the collection of minor forest produce such as yams, various leaves, roots, barks and herbs used in medicines and tamarind, etc. Finally, the comparison of the economic life of both the villages is presented.

The fourth chapter is divided into two parts. In part I, an effort is made to present, at the macro level, how in India changes occurred in the ecology and economy in the tribal habitat due to urbanization and commercialization process, initiated by colonial rulers and the manner in which it affected tribal people in general due to Government of India's forest policies and introduction of reservation of forest or restrictions for exploitation of forest resources.

In part II, how the Paliyans in particular were affected by such restrictions and lost their traditional access to resources has been discussed. How they have adapted to new economic pursuits due to such changes and are availing new opportunities have also been presented.

As in the previous chapter, here also data are presented at two levels; at the macro and micro levels. At first, the tribal development policies and their implementation are discussed. Also the constitutional safeguards for tribal development and the different projects and their implementation in different five year plans are discussed. In the second part, the developmental processes in the two villages are discussed. Finally, the tribal development and national welfare at the macro level and their impact at the micro level are analysed along with the evaluation of different projects in the third part.

In this chapter, analysis has been made of different situation starting from pre-colonial days to post-independence era for Paliyan's development along with reasons of success and failure. Here the analysis of different hypotheses and their results has been discussed further. An effort has been made to identify the problems of development for the Paliyans and how such problems could be solved or eradicated with a view to implement successful tribal development policies in India.

Finally, with specific reference to the Paliyan situation, an effort has been made to analyse the paradoxical situations in tribal developmental policies and their implementation. A few suggestions have also been given at the end, focusing how to eliminate such situations on the basis of the analysis of the traditional socioeconomic situation and the traditional value system with the present trend of change.

With a view to abolish bonded labour system, a serious effort was made in the village of the Kamarajar District. Multi-faceted developmental projects were also adopted in this traditional village. The same is taken in this monograph for a detailed case study. The factors responsible for failure have been discussed in this volume.

This volume is an outcome of field work during my fellowship tenure with Anthropological Survey of India during which I took up my Ph.D. I am deeply grateful to Professor Dr. P. Govinda Reddy for motivating me for studies in Anthropology and to **Anthropological Survey of India** for support at different levels.

The idea of taking up Paliyan tribe was suggested to me by my father Late (Prof.) Uma Charan Mohanty as no intensive work was conducted by any Indians on this community and his consistent guidance, vision, correction of draft and encouragement at different level enabled me to complete the work successfully . The sympathy of my former Directors especially Dr. R.K. Bhattacharya during crucial hours is note worthy. The consistent persuasion and suggestion of Dr. A. Sarkar and my brother- in-law Mr. Aswini Das to bring out this volume in a short spell of time is heartily acknowledged. I also express my hearty gratitude Professor (Dr). Jagannath Dash for his valuable comments on the volume.

I am highly indebted to St. Jesuit Society of Tamilnadu and to Mr. Kondeyan and Gopal for their help during the field work the result of which is placed in the book.

In this great effort there has been constant support of my present Director Shri G.S.Rautela, Dr.(Mrs)K.Chakrabarty Chairperson P&P and Mr.D N.Pandey Head of Office. I extend my sincere thanks to them

This volume could not have been complete without enthusiastic cooperation of my colleague Mr. Y. Vijay Kumar. Grateful thanks are due to Mr. B.K. Sahu, Mr. B.B. Dash for their timely support for the completion of this volume.

**Bibhu K. Mohanty**

Bhubaneswar

# Abbrevation

| | | |
|---|---|---|
| 1. | D.R.D.A. | District Rural Development Agency. |
| 2. | I.R.D.P. | Integrated Rural Development programme. |
| 3. | I.T.D.P. | Integrated Tribal Development Project. |
| 4. | J.R.Y. | Jawahar Rojagar Yojana. |
| 5. | M.F.D.A. | Marginal Farmers and Agricultural Labourers Development Agencies. |
| 6. | M.A.D.A. | Modified Area Development Approach. |
| 7. | M.N.P. | Minimum Needs Programme. |
| 8. | M.T.D.P. | Multipurpose Tribal Development Projects. |
| 9. | N.G.O. | Non Government Organisation |
| 10. | S.C. | Scheduled Caste |
| 11. | S.C.A. | Special Central Assistance |
| 12. | S.T. | Scheduled Tribe |
| 13. | S.F.D.A. | Small Farmers Development Agencies. |
| 14. | S.O.C.O. | Society for Community Organization Trust. |
| 15. | T.D.A. | Tribal Development Agencies. |
| 16. | T.D.B. | Tribal Development Block. |

# List of Tables

# List of Charts



# List of Map



# List of Pictures

# List of Photographs

# Introduction

Anthropology developed its theoretical postulates and methodological strategies through the study of primitive communities. Nevertheless, in course of time, anthropologists have become more interested in expanding its scope and have thus refined their methodological approaches for the study of more sophisticated peasant and urban societies.

At present, with the expansion of science and technology and the spread of monetized economy and commercial enterprises, the modernization process has entered into the remote corners of the world. Yet there still exist a number of sparsely populated, very small communities in different secluded pockets of the world, which are in the process of rapid transformation or extinction. Recently, such communities of hunters and gatherers have drawn the attention of a fairly good number of anthropologists, even at the global level.

In the highly sophisticated modern world today, the process of industrial advancement based on the high technology is destroying or threatening to destroy the ecological balance paving the way to environmental hazards. Hence in the present context, ecological studies are assuming great significance. Scientists and social researchers are now deeply interested to make in-depth studies of primitive communities with a view to learn how they managed their ecological resources, the manner in which they developed their eco-friendly productive processes, and how they adapted themselves to the environmental situation of their habitat.

On the other hand, due to technological development coupled with reckless commercial exploitation of natural resources, underdeveloped countries remain victims of poverty, unemployment and ill health. Thus economic disparity is creating rich and poor nations.

## The Problem of Study

In a country like India, which contains the largest number of diverse castes, tribes and communities that represent different stages of development from Paleolithic stage to modern nuclear age, bringing unity and solving their problems of development is a Herculean task. The backward tribal communities are at the lower rungs of Indian society. Since Independence, inspite of ceaseless efforts by Central and State Governments, the problems of backward tribal communities still remain unsolved. The problems of small scattered food gathering communities especially have escaped the attention of Government of India.

Recently, the Government of India has identified a few such small foraging, isolated, poorly literate tribal communities, and included them under the newly categorized list of primitive tribes. Yet, unfortunately there are still some such communities, which could not be included due to some reason or other and escaped the attention of the government. Even of such communities which are enlisted as primitive and for whom special schemes have been formulated for their upliftment, nothing significant has been achieved so far, for their emancipation from age old problems of poverty ill-health and illiteracy. Further an important question can be raised: when Government is making special projects for their upliftment, why no sustained development is achieved for these nomadic or semi nomadic communities who are supposed to have occupied the lowest rung in the social hierarchy?

Hunters and gatherers or the foraging communities lead a nomadic or semi-nomadic band life in an extensive, isolated sparsely populated, natural environment. Such primitive tribes, irrespective of their identification and declaration by the Government of India as such, should be given prime importance in the national planning agenda.

Further, foraging communities all over the world are facing rapid and tremendous change due to modernization process which affects the ecology of their habitats and in transforming their economic life in a significant way. The hunters and gatherers or foraging communities have their distinctive way of life, which is under threat due to ecological hazards that they face in the rapidly changing world. The study of such communities should get topmost priority in view of the fact that their cultural adaptations are likely to vanish rapidly in the present context.

Basing on the above stated need, the Paliyans have been identified for an empirical study. With a view to show how a self-sufficient, small tribal community has lost its economic independence due to changing ecology and economy, introduced by colonial rulers, Paliyans provide an excellent model. Further an effort was also made to assess to what extent developmental processes initiated since independence by both the Government and non-Governmental agencies are achieving a degree of success or meeting failure in their efforts.

In the present work, an effort is also being made to explore and present the traditional culture of the Paliyans, highlighting their self-sufficient economy and their complete dependence on nature. Further, an attempt has been made to show how they changed slowly from their self-sufficient life in the state of nature to their position as labourer during the colonial period. It is presumed that the Paliyans, in the beginning of colonial rule were living in the free state of nature, enjoying economic selfsufficiency and socio-political freedom. But gradually the tribe had to undergo the process of subordination to the forest officials, forest contractors, police and other Government officials. Local traders, business men and surrounding caste people maintained good social and economic relationship with such officials of the Government and kept the Paliyans in their clutches. Traders or forest contractors utilized the Paliyans labour for their own advantages. Of course such subordination of the Paliyans has been a long process covering a period of roughly one hundred years. With their deteriorating status, the Paliyans have been exploited by various agencies or groups and continued to live at a subsistence level. Fortunately, now there are signs of their socio-economic improvement due to developmental efforts by the government and non-governmental agencies. It is difficult to evaluate their developmental success quantitatively at the present situation. However, we try to first identify the problems of the Paliyans as existing at present.

Before the colonial period, the Paliyans are presumed to have been the dominant group and the migrant caste people must have depended on them and thereby they might have enjoyed fairly high status. But during the colonial rule, when more of plains people migrated to areas inhabited by Paliyans, gradually their status declined. Since their customs differed from local Hindu customs, they were ridiculed by members of dominant groups as a result of which they developed inferiority complex in their interaction with the outside world.

During the colonial period, the Paliyans underwent a transitional phase from full freedom in a state of nature to a life of hardship under the control of rich traders or forest contractors belonging to higher castes or classes under whose direction they had to work hard for their subsistence. Since the Paliyans were not acquainted thoroughly with monetized economy they became indebted and soon became bonded labourers, by their neighbours. Due to trickery of their employers, they soon became addicted to liquor, ganja and other such narcotic items. More recently they became victims of allurements of modern days, such as films (especially blue films) and other temptations. Some of the Paliyan youths have been observed to be drifting away from their traditional norms and value systems.

## Objectives of the Present Study

1. To explore the traditional culture of the Paliyans and analyse to what extent their culture was influenced by their environment.
2. To study the socio-economic changes of the Paliyans from food gathering stage to agricultural labourer or to the adoption of modern mixed economy.
3. To study the nature of exploitation that they suffered from in the past and are still suffering to some extent under the traders, forest officials and other surrounding people who employ them for their own financial benefits.
4. To examine and evaluate the developmental strategies adopted to improve the socio-economic condition of the Paliyans by different agencies viz, Government organizations, Christian missionaries, and Voluntary organizations, etc.
5. To explore and identify the major obstacles to the development of the Paliyans.
6. To explore the loopholes in the existing policies for tribal development.
7. To explore to what extent the Paliyans are developing at present by adopting new opportunities.
8. To explore what development strategies are needed to be adopted by Government and other welfare agencies.

In order to fulfill the above objectives the following theoretical perspectives have been utilized for the collection and analysis of data and the presentation of the same in this research work.

## THEORETICAL PERSPECTIVE

### Cultural ecology

Though the term environment and ecology are used some times interchangeably the latter is much wider. Environment means "the complex of climatic, edaphic and biotic factors that act upon an organism. The term, 'ecology', in a broader sense refers to 'the totality of relations' between environment and organism, behavioral aspect may be absent with the use of environment while ecology does include behavioral pattern of the organism" (Srivastava, A.R.N: 1990:279).

Prior to the development of the concept of cultural ecology, geographers like Ratzel and his followers developed the theory of "Environmental determinism". The concept environment on the culture of a given society.

The general orientation of environmental explanations in anthropology shifted from determinism to possibilism in 1920s and 1930s. Possibilism emphasizes that historical tradition rather than environment resulted in the origin and continuity of specific cultural features and patterns in a given society. Thus this approach has been called as "historical possibilism". Which emphasizes that nature offers people with a certain number of possibilities but the historical and cultural factors direct people regarding the possibility that should actually be chosen. According to this view culture is environmentally selective. The deterministic view holds that the environment actively shapes man and culture and the possibilistic view on the contrary asserts that environment is a limiting factor.

Environmental determinism was strongly opposed by Kroeber, Dyril Forde and others. Kroeber emphasized the "historical and cultural influence" as determinants of cultural phenomena. C. Daryall Forde, the author of a widely read book. Habitat Economy and Society, "found no simple correlations or generalization existing between environment and society". (SRIVASTAVA, A.R.N.: 1990-280).

## Emergence of Stewards cultural ecology

Julian H. Steward wrote as early as 1930s, on the ecological point of view in anthropology. The most important contribution of Steward was his method of "Cultural ecology". This emphasized that environment and culture are not separate spheres but are interdependent and involved in "— dialect interplay—or what is called feed back or reciprocal causality —" (Kapaln and Manners: 1972:79).

The main thrust of cultural ecology is to explore the intimate relationship between culture and environment and such intimate interaction is considered natural, Steward first introduced the concept of "adaptation" in the study of the cultural behaviour and demonstrated it as his contribution to cultural ecology. He laid emphasis on interactional analysis of the relationship between environment and culture. But the "adaptive interaction" as a notion which forms the basis of cultural ecology was put forth by Steward in 1955. According to him it is culture that plays an important role at certain times and environment at certain others. He believed that some aspects of culture are subjected to a strong environmental relationship as compared to other aspects.

Ecological analysis is then used to explain cross-cultural similarities in the "culture core". The "culture core" constitutes the economic sector of the society as the base and those features that are "most closely related to subsistence activities and economic arrangements", are the resultants of the economic activity." The core includes such social, political and religious patterns as are empirically determined to be closely connected with these arrangements." "Cultural ecology pays primary attention to those features which empirical analysis shows to be most closely involved in the utilization of environment in culturally prescribed ways". (Steward: 1955:37).

**According to Steward the method of cultural ecology evolved**

1. The analysis of the interaction between the environment and technology that a group uses. Same environmental features may be more crucial than other in influencing technology (culture).
2. Analysis of the pattern of activity required by the exploitative technology used in given area. Does the technology require co-operation? Does cooperation result in more efficient utilization of resources? Therefore, to Steward the important point is "adaptive strategy of the group in a given environment". (Steward 1955: 41).
3. Description of how the behaviours involved in exploiting the environment are affected by other aspects of the culture. The organization of Kin groups socio-political organizations, population density, material cultures are some of the areas of the social cultural life that may be involved". (Srivastav, A.R.N.: 1990:282).

Steward by prescribing the above fundamental procedures, tried to demonstrate his concept of 'cultural ecology' approach, which is considered more scientific in studies that deal with environment and cultural relationships.

Subsequently many anthropologists have adopted the methods of cultural ecology as prescribed by Steward yet: a good number of anthropologists have tried to improve this approach on the basis of their field investigation. Among such people Marshal D shalins (1969) studied ecology from evolutionary point of view. Frederick Barth's contribution to cultural ecology is very valuable. In his book "Ecological relationship of ethnic groups in Swat, Northern Pakistan"(1969) he has provided new insights into cultural ecology. While studying the environmental and cultural relations, he formulated a frame work which deals with two or more cultures co-existing in the same ecological zone and the process of adoption born at the culture as well as the environmental levels. By adopting the concept of "Ecological Niche", he discusses in detail how the peaceful co-existence of people belonging to different cultural backgrounds with different occupations but living in the same geographical conditions is possible. His study throws light on how natural resources are utilized by different ethnic groups through minimum conflict. Though different ethnic groups may occupy and share the same natural environment and resources, they establish distinct niches in the natural environment and prevent competition for resources. They are linked through, a kind of reciprocal exchange of goods and services. Thus a kind of close inter dependence exists among the groups which ultimately leads to a kind of smooth and harmonious co-existence of populations in the same environmental conditions but in separate niches.

## Ecological studies in India

Ecological studies in India started as early as 1930, when R.K. Mukherjee published his pioneering paer "Ecological contributions to sociology". He also wrote the book 'Social Ecology' (1942). From the beginning, anthropologists in their monographs on Indian tribes have been writing on the environmental description of the habitat of particular tribe. Subsequently with the publication of Julian Stewards "Cultural Ecology" in his famous book 'Culture Change' anthropologists working on Indian tribes were influenced by Stewards theoretical approach and tried to apply it in their analysis of the role of environment in the formation of the culture of the particular tribe. Vidyarthis monograph entitled "The Malers: Nature Man Spirit complex" (1963) is one of such monograph on an Indian tribe. Fortunately Steward himself has given an enlightening foreword to that monograph. Subsequently a good number of monographs on Indian tribes appeared and such monographs are being published from time to time. Besides a good number of Ph.D. thesis are being written in different universities in which the theory of cultural ecology is the main thrust of the theoretical analysis. The ecological studies have become so popular in India that a number of seminars, workshops are organized from time to time. Such papers are published in anthropological journals and some times such papers are published as a special issue of a particular journal. Recently a good number of valuable published papers relating to India, on ecology have been collected and edited by Guha (1995) and this has been published under the title "Social Ecology".

## Ecology and development studies in India

A new trend is emerging relating to ecological studies in India. With the advancement of developmental activities, there are large-scale destruction of forest wealth and other resources. Thus the developmental process such as construction of roads multi-purpose river projects and opening of urban centres in isolated tribal habitat has raised the question. What should take primacy? It is wondered whether development of the tribal communities of a particular area is to be prioritized or to fulfill the so-called interest of the nation by sacrificing natural resources of that area. As a consequence large-scale social movements have been generated such as 'Chipco movement'. 'Sardar Sarobar Bachao' etc. Based on such mass movements in a democratic country, with its welfare ideology, developmental ecology is slowly resources of India and the lost of ecological balance which affects tribal and rural communities anthropologist, sociologist and other social scientists have generated more prospective write-ups and monographs focusing on current developmental projects and the ecological hazards that are emerging out of such efforts. Such papers which are contained in Vandna Siva's edited book, "Ecology and Politics of Survival: Conflict over Natural Resources in India" (1991) is a pioneering book on Developmental Ecology. The book of Madhav Gadgil and Rama Chandra Guha's (1992), "This fissured land: And ecological history of India" has also brought a new awareness for protection of forests and other valuable resources.

Gadgil and Malhotra (1994) with insight from Irawati Karve seem to suggest that each by inheriting an occupational space also acquires an ecological space. The allottees of these ecological spaces develop their own sustainable patterns of use of natural resources. They theorise that changes brought by contemporary market forces have destroyed this harmony.

Thus ecological studies in India is in no way less productive and through the papers relating to developmental projects and destructions of natural resources a new area is emerging dealing with "Ecology and Development".

## Recent studies on hunters and gatherers

Cultural Man has been on earth for some 2000000 years; for over 99 percent of this period he has lived as a hunter gatherer. (Lee Richard B. and Irven Devore 1968). But the present day Hunters and Gatherers are disappearing very rapidly. On the other hand research on Hunters and Gatherers are increasing rapidly since 1965 onwards. Researchers are now concentrating on the study of ecology, social structure, gender role demographic, techno-economic and ritual practices of different groups of Hunters and food-gatherers scattered in different isolated pockets of the world. Through recent researches some classification of Hunter-Gatherers have emerged and some concepts have been developed. A few concepts such as "Immediate return societies versus Delayed return societies", "Simple Hunter Gatherers versus Complex Hunters Gatherers", "Affluent Hunters Gatherers", "Social stratification", "Culture contact situation", Governmental intervention for the development of hunter Gatherers, are some of the key issues in present day research on Hunters and gatherers.

In present situation in India, two issues, the "Culture Contact situation" and "Government Intervention" are important issues. Both traditional process of acculturation and the recent process of acculturation are two important processes for the transformation of Gatherer Hunters societies of India. Both these issues have been discussed in the present volume.

## Status of present day gatherer hunters of India

The three terms Hunter – gatherers, Gatherer – Hunters and foragers which include both the groups, almost convey the same meaning. But in Hunter – Gatherer, hunting is the dominant occupation while in the term Gatherer-Hunter, gathering takes precedence over hunting. Therefore in Indian context all the foragers should be termed as Gatherer-Hunters or simply as Food Gatherers.

But in recent years when research relating to Hunters – Gatherers is gaining importance, the debate is also going on as to the status of Hunters Gatherers of India. Can they be considered same or similar as in other parts of the world such as Hunter – Gatherers of Africa, Australia or Hunter Gatherers of North and South America.

In this connection a perspective paper has been written by Richard G. Fox (1969) entitled, "Professional Primitives: Hunters and Gatherers of nuclear South Asia". The author has taken five Hunters and Gatherers of South Asia. Viz; Kadar, Birhor, Chenchu, Vedda, Nayadi. He has presented the social ecology and social organization of each of such tribes. He has put forth three points as detailed below.

1. "The spatial separation of hunters and –gathers from Hindu culture and society has been exaggerated———"
2. Rather than being independent, primitive fossils, Indian – hunters and gatherers represent occupationally specialized productive untis similar to caste groups———"
3. The collection of forest products for external barter or exchange has various effects on the internal social organization of South Asian hunters and Gatherers (Fox, R.G: 1969: 141 – 142).

Fox, concludes, "from this perspective, Indian hunting-and gathering groups are viewed as marginal economic specialists for traditional Indian civilization. (Fox, Richard. G: 1969: 141 – 142).

## Anthropological approach to tribal development

### *Applied Anthropology*

During the colonial period, British administration took keen interest to collect data on tribal communities and later they developed a tribal policy for well being and upliftment of tribal communities. The present tribal policy in free India is built upon the foundation of policies developed during British period. Only during British raj applied anthropology emanated as an applied discipline out of the anthropological theories. Applied Anthropologists during the colonial period, in order to maintain scientific neutrality kept themselves free from policy formulation or execution of Government plans and projects. During the colonial rule, anthropologists developed the theories based on data collected scientifically. During the same period eminent British anthropologists viz; Malinowski, Nadel, Radcliffe Brown and Raymond Firth etc. contributed theories relating to applied anthropology.

### *Action Anthropology*

In field situation Sol Tax and his students developed the theory of Action Anthropology. Applied Anthropologists, in order to maintain scientific neutrality avoid making any intervention in the culture of the target group and leave such work to administrators. Action anthropologists follow the clinical method and do not hesitate to interfere in the affairs of a community for their development with a view to solve the problems of a group of people who need such assistance or help. Sol Tax advised to help such distressed people but he has suggested certain principles to be followed which would maintain the scientific neutrality along with field work and community service for such disadvantaged group through action anthropology. According to him action anthropologists give propositions which can be applied universally.

1. The value of truth in which we believe that truth & knowledge are more constructive in the long run than falsehood & superstition.
2. We feel more strongly the value of freedom. Freedom in our context means usually freedom for individuals to choose the group with which to identify, & freedom for a community to choose its way of life.

3. A third value – or is it a principle of operation? – is a kind of law of parsimony which tells us not to settle questions of values unless they concerns us. [Sol tax 1968:37].

After independence of India and formation of the largest democratic welfare state, Government of India tried to remove poverty and to achieve development in all spheres of life such as health, education etc & tried to achieve higher quality of life. In this regard various theoretical approaches have been developed by anthropologists which can be used as guide lines in the developmental process in various situations. In the framework of nation state, economists dominated the scene of development. Even the former colonies have followed such pattern. So, developmental programmes were cast in the economic mode. And one of such reasons could be the agenda associated with this discipline in contrast to other disciplines.

"Economic as a profession has a social agenda: to develop the economy not merely to collect economic data but to construct theories. Sociology has something of an agenda to improve society, but it is vague and no one quite knows how to do it. Anthropology has no historic agenda its task was pure research the construction of cultural portraits" (Bennet John W: 1989).

Under these circumstances role of economist was sought, as they were thought to be the right professional to guide about economic development. But Mandelbaum (1977) the anthropologist agrees that in the initial stage, role of economist was necessary as hundreds of people were living and live in dire poverty. So the slogan that "change the economy and the rest follows" was so much in operation that parameters of social development such as education migration and labour are considered economic and thus the cost of gains of proposed changes are measured in terms of money. Despite these policies it was still evident that though material production had gone up yet there are many who have not benefited from these changes.

For example smaller communities of the country like pastorialist, gatherer hunter etc. are very likely to be ignored by the general developmental programmes conceived by economists. So it was felt that to reach different section of people widening of developmental policies was necessary. And to do this it was felt that knowledge of social and behavioural sciences could be tapped.

### *Role of Anthropology in developmental strategy*

But then we will limit our discussion on anthropologists only due to paucity of space. Anthropologist's emphasis on holistic view disposes them, to take into account the social matrix in which particular problems are embedded. Their practice of fieldwork directs attention to development as perceived from within, how the people understand, accept, or reject introduced innovations. Anthropologist's are also inclined to relate what they learn in the micro view to the macro view to some general perspectives on the human career. Comparative studies done by anthropologists have often taken a wide range sometimes a global span & thus it suits the global forces that increasingly affect the peoples of modern world.

Mandelbaum writes, one broad task of social anthropologists is to study the possible sequences of sociocultural change open for a particular people at a particular time and to make recommendations for policy and action. Development studies by anthropologists have dealt, for the most part, with much narrower and more specific matters than the grandeur issues of national policy alternatives. (Mandlebaum, D.G. - 1977).

It has been recognized that many developmental programmes have failed miserably as they were sociologically and anthropologically ill informed and ill conceived. So more and more socio-anthropological knowledge were used in project planning which began in 1970.

During this phase an attempt was made to move away from trickling down theories to poverty alleviation through development which was proposed and initiated. Here the concept of "Target group" (those who are poor with an income below the absolute or relative threshold of poverty) was the main emphasis. Cernea (1985) while emphasizing on this transition writes.

"The shift from a virtually exclusive emphasis on physical infrastructure to a recognition of social structure, and from free swinging ethnocentricity to the recognition of indigenous cultures was part of the shift in the direction of actor (people) centered development projects".

In brief, to bring successful development of the target group, anthropologists try to study carefully the culture and social structure. The culture includes the norms, values and traditional customs or institutions. The social structure includes the different groups within a community and their intra and intergroup relationship. With such detailed studies of culture and social structure, anthropologists try to identify the stimulants and barriers in a developmental process and with such detailed information try to help successful adoption of innovations that are likely to bring sustained development. Some adopt action anthropology or participatory research and through participation in developmental process they try to carefully analyse the stages of progress. With detailed study and research of developmental process, gradually anthropolotgists are gaining experience and insights to tackle problems relating to development. Thus they are better equipped to accelerate the process of development and contribute in policy formulation, and in monitoring and evaluating the developmental projects. While analyzing the developmental process they try to identify the chained reactions after the introduction of the innovations, which are supposed to bring development to the target group, and also they try to visualize what would be the unintended consequences of such developmental process. Thus they try to caution the planners and the beneficiaries as regards unintended consequences that should be avoided or eliminated.

## METHODOLOGICAL PERSPECTIVE

The present study tries to adopt exploratory research design since detailed information are lacking as regards its history and factual information about various aspects of its culture and social structure. The few data that are available at present does not give the true picture. For collection of data, both survey and case study methods has been used. In certain areas, such as census, collection of economic data, survey method is being used. In other areas such as collection of life history, community life or band organization, life cycle rituals and political organization, study of conflict and cooperation etc, case study method has been adopted. As regards techniques of data collection, traditional methods such as genealogical method, structured and unstructured interview, participant and non-participant observations have been utilized. Case study method has been utilized very often when adequate factual data could not be gathered directly.

### Stages of Field Work & Data Collection

Before starting the data collection the author tried to utilize secondary sources such as reports contained in gazetteers, census reports and ethnographic accounts published in monographs and scientific journals. Further the records of different government offices and documents maintained by voluntary organizations and archives maintained by missionary organizations and government were also utilized for collection of reliable data.

Next the focus of the author was on the distribution of Paliyans. For this the author took up reconnoiter visits and found that Paliyans are mainly distributed in following three areas.

1. (a) Palani hills and Sirumalai hills of Western ghats located in Dindigal district. The highest concentration of Paliyans is in this area of Palani hills.
   (b) Varshanad valley and foot hills of Western ghats in the Madurai District.

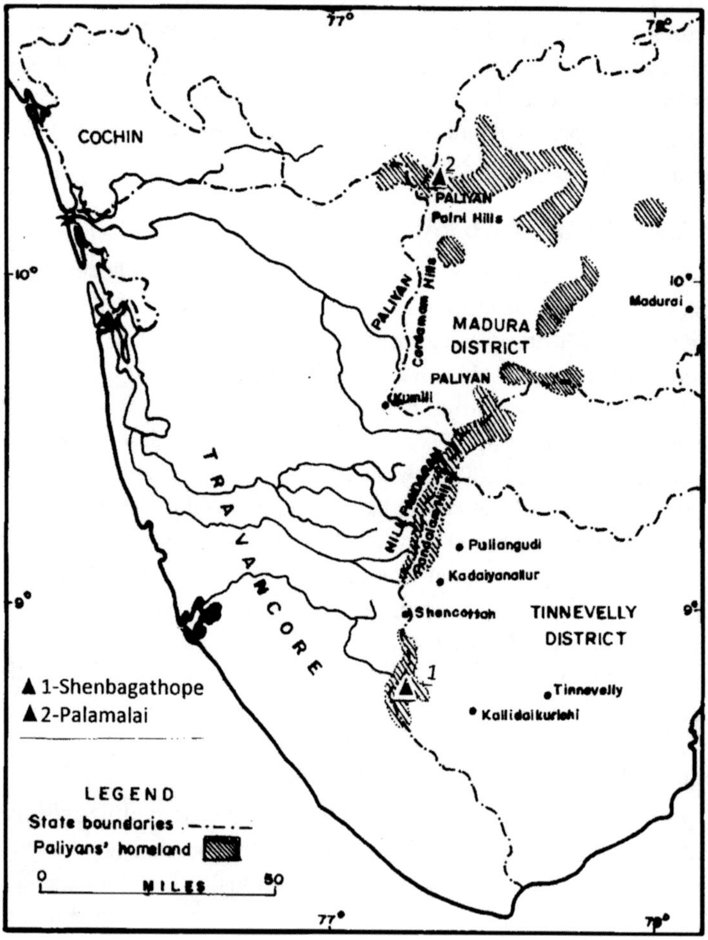


2. The Srivilliputur Taluk and Watrap Talluk of Kamarajar district which contains scattered hamlets of the Paliyans at the foot of the hills such as Shenbagathope of Srivilliputur area, Aiyanar Koil of Rajapalayam area are the next important concentration of the Paliyans.

3. The Tirunelvelli district containing Sankaran Koil and other acculturised hamlets of the sparsely populated Palliyan hamlets is the next area of dispersed Paliyan area of distribution.

But the third area of Paliyan population is so sparse and since majority of them have been converted to Christianity the author dropped the third area of distribution of the Paliyans for his field work. Their number is insignificant and they have lost their traditional way of life.

## Selection of two sample villages for intensive study

Out of the two above mentioned Paliyan areas, one village from each area has been selected for intensive study. While selecting the two villages criteria such as communication facilities, exposure to outside influences etc. have been taken into consideration as these factors contribute towards accelerated developmental process. Two villages from two areas helped in making the comparative analysis. Shenbagathope village was selected from Srivilliputur Taluk and Palamalai village was selected from Kodaikanal area of Palani hills.

## Difficulties faced and Establishment of Rapport

(a) Language was the main obstacle in carrying out the field work. Tamil was not the mother tongue and therefore the author had to take a long time to master the language so as to collect reliable data directly from the informants. Gaining mastery over the language so as to gain confidence of informants took quite a long time.

(b) Most of the Paliyans are very suspicious of the outsiders and therefore did not cooperate with the author at the initial stage. It took quite a long time to gain their confidence.

(c) There were some problems in collecting accurate economic data. Specially majority of informants hesitted to give accurate data on their loans and other expenditure incurred during consumption of alcoholic drinks and narcotics, etc. Specially those who depend on food gathering economy could not give any specific data regarding income and expenditure. Anthropological method of collecting data through observation was also time consuming affair. Information on day to day income and expenditure was relatively more forthcoming in Palamalai (of Palani hills) than in the Shenbagathope of Srivilliputhur area. Nevertheless efforts were taken to get more or less accurate informations by interviewing petty shop keepers, neighbouring castes, and outsiders who employ the Paliyans as their labourers. Cross checking those data with the statement of the Paliyans, some realistic picture could emerge about the present economic situation of the Paliyans.

(d) Details of developmental activities and financial expenditure for each project and the degree of success of such projects could not be assessed as the concerned officials did not fully cooperate with the researcher. They hesitated to give details of information regarding development to avoid contradictions and exposing their own lacunas.

## Working Hypotheses of the Present Study

1. During the early period of the colonial rule, the Paliyans had already reached the status of caste (Jati) through their close interaction with the neighbouring castes and communities. A caste has three important features (i) Endogamy (ii) Traditional occupation (iii) Caste Panchayat which regulates the behaviour of the members of the caste. Before the beginning of colonial rule they had already adopted endogamy, foraging was their exclusive traditional occupation and the leader of the village, who acts with others can be treated as the caste Panchayat.
2. The Paliyans as gatherers and hunters differ substantially from the hunters and gatherers found in different parts of the world, because of different environmental and cultural set ups they live in.
3. Commodifiation of forests by the British was responsible for the low status and poverty of the Paliyans.
4. The Paliyans who depended mostly on the collection of minor forest produce, became virtually victims of government servants, forest contractors and traders. Without changing the exploitative system and the forest policy, welfare of the Paliyans is not possible.
5. Changing ecology and introduction of monetized economy were responsible for the poverty of the Paliyans.
6. People of upper strata, maintained their social nexus with the government machinery and with their support kept the Paliyans subordinate to them.
7. Welfare schemes for the upliftment of the foraging Paliyans could not be successful unless these people were motivated to participate in their own welfare programmes for a higher and better quality of life.

# 1

# A Profile of the Traditional culture

## SECTION – I

### GENERAL INTRODUCTION

During British rule in India, a large number of ethnographic works on tribes, castes and communities, were generated through the encouragement and blessings of British administration. For the first time in the long history of Indian civilization, the attention of administration was particularly focused on the tribal communities who live in hills and forest, in inhospitable climate. Moved by curiosity and impressed by the diversities and antiquities of such cultures, travelers of foreign countries, Christian missionaries, philanthropic administrators and scholars of both India and western countries also contributed research papers, reports and notes either brief or elaborate and description of their experience as regards tribal people and their culture. No doubt, some times, such descriptive accounts were not based on scientific investigation and lacked any theoretical orientation. Yet these accounts help us to visualize those by-gone days when communication was not that easy and the modern education and civilization had not entered the tribal pockets. However brief such accounts may be, it provides us the base line data for analysis of the changing tribal culture which is undergoing transformation under the impact of modernization and developmental activities initiated and advanced in the post independence era by Indian Government, as a welfare state. The present chapter is based on the earlier documented ethnographic accounts relating to Paliyans, a tribe of Tamil Nadu, so that the developmental process in this tiny yet scattered community, can be analysed in proper perspective, while focusing the developmental achievements or failures as a result of welfare measures ushered in by the government of India and the State Government. Thus the presentation of the traditional culture of the Paliyans has three objectives in view.

## A. Objectives of studying the traditional Paliyan culture

1. The concept "tribe" is most anomalous and leads to a great confusion in free India. The description of earlier reporters who had direct encounter with the tribal communities, not only give us scope to witness the glimpses of traditional culture unadulterated by outsiders, but also it provides to some extent, what features of a community were considered as tribal or primitive in those earlier days.

   Further recently "Hunters and Food-gatherers", as a special category of primitive communities living in particular ecological niches have drawn the attention of anthropologists. The description of the traditional culture of the Paliyans which is undergoing the process of rapid change, may be useful in analyzing the nature of their culture and interaction with outsiders to understand the tribal life of India in more specific way.
2. At present tribals are changing rapidly after the independence of India and the Paliyans, a tribal community, is no way excluded from such contemporary phenomena. To assess the process of change occurring at present amongst the Paliyans, the earlier documents, may serve as a rough base line for analyzing their traditional culture, social structure and their transformation.
3. In the present developmental process, such features of the traditional Paliyan culture which inhibit the process of change, or such features which help in identifying the felt needs of the community may go a long way in accelerating the process of change and in directing it towards the development of the community. Further such documents may help us to understand their basic problems which were also existing in those early days.

## (b) Distribution and the divisions of the tribal communities in peninsular India

"Guha" has broadly categorized three regions of India in which most of the tribes of India are concentrated. Out of such regions peninsular India contains a good number of very small tribes dispersed widely. The general features of these tribes have been briefly described by Guha as given below.

1. Excepting the Toda, Badagas and Kotas of the Nilgiri hills the basis of life of these aborigines centers around hunting and food gathering as they have not yet developed a settled community life, but wander from one place to another in quest of food.

2. Their social structure is based on a dual organization with authority vested in the head man who settles disputes and perform the rituals of hunt.
3. They now speak corrupt forms of the Dravidian languages.
4. Physically these are of short and medium statures deep chocolate brown in colour, and with broad flat noses and thick lips. (Guha: 1973:29).

J.H. Hutton in describing such most primitive of the south Indian forest tribes gives the accounts of the Kadar of Cochin, Mala-pandaram of Travancore and Paliyans of Madura districts. Hutton considers both Mala-Pandaram and Paliyans as similar. Luiz also observes more emphatically that the Mala-Pandaram is only a section of the Paliyans. Since we are more concerned with Paliyans we would rather concentrate on this tribe henceforth.

The Paliyan is a small south Indian tribe which is now sparsely distributed in the states of Tamil Nadu and Kerala. In Tamil Nadu, in the Dindigul, Madurai, Kamarajar and Tirunelveli districts, Paliyans are sparsely distributed.

Tribes of Peninsular India can be broadly divided into four groups on the basis of economic pursuits (1) Hunters and food gatherers (2) Pastoralists and artisans (3) Shifting cultivators (4) Settled cultivators and/or agricultural labourer. The Paliyans belong to the first category. In the present chapter an effort is made to portray a profile of paliyan culture based on documents of the Pre-Independence era which are mostly ethnographic accounts or brief notes published in Government reports such as Gazetteers, Manuals or papers published in scientific journals. Such documents try to cover the whole tribe, or community and to cover all aspects of their life. The following pages will provide such all-inclusive descriptive accounts to give a complete picture of Paliyan society and culture. Of course, in post independence era though such trend is still continuing in district Gazetteers (in the peoples chapters) and in census reports, in the present situation more scientific papers and theorctically oriented monographs prepared by scientifically trained scholars are becoming more conspicuous. The works of the post-Independence era will be referred at relevant situations in subsequent chapters, along with data collected through empirical field work.

### (c) Distribution of Paliyans

The Paliyans are distributed both in Tamil Nadu & Kerala. The distribution of Paliyan is done on taking into account the three categories namely Paliyan, Palliyan and Paliyar as a single category. This is based on the 1961 census ethnographic note which states that:

> The name of the tribe is known locally as Paliyan and not as Palliyan or Paliyar as indicated in the list of scheduled tribes notified by the President of India
>
> (**NAMBIAR**: 1961)

L.A. Krishna lyer writes that "Paliyans of Vandanmet state that they formerly lived in Nelliyani Paliyadi & Tamarakulam in Gudalur of Madura District, when a Pandaram told them that they would find it congenial to go and live in Cardamom hills in Travancore which were uninhabited and they accordingly came and settled down at Vandanment etc. (lyer L.A.K.: 1937:9). There are reports of 19th & early 20th century which narrates Paliyan rove in small parties through the jungle clad gorges that fringe the Upper Palani Plateau. Though an exhaustive reconnoiter tour was conducted yet it was not possible to cover all the hamlets in every district due to the scattered nature of their distribution. Paliyans cover an area of 150 by 80 miles area. So the researcher has to depend upon census record in the analysis of his data. Again the focus would be on the Paliyans inhabiting in the Tamil Nadu side which forms the major chunk of this tribe. Apart from Tamil Nadu they are also found in the neighbouring state of Kerala inhabiting in the same ecological and geographical contiguous region. Another noteworthy features is that their habitat is always associated with the scenic mountains of Western Ghats, which forms the life line for them and the significance of this will be discussed later. Their main concentration can be found in the following areas of Tamil Nadu.

1. Palani hills
2. Sirumalai hills
3. Varshanad hills of Madurai
4. Srivilliputhur areas of present Kamarajar district
5. A few Paliyans of Tirunelvelli district.

## SECTION – II

## TRADITIONAL PALIYAN CULTURE

The first 'Manual' of the district of Madura was 'The Madura Country' of J.H. Nelson. I.C.S published at Madras in 1868. The subsequent manual was the Madura Manual-1891 of Western Francis which was based on Nelson's account. From the Madura Country of 1891 the following extract is given below to show the nature of such brief notes meant for administrators.

## (a) Extract from the Madura Manual – 1891 and its analysis

The Paliyans are another mountain caste but of a far lower type of humanity than the Poleiyans. In fact they are savages and decline to adopt the most simple usage's of ordinary men, having neither houses, clothes nor any kind of property. Roaming over the hills, they satisfy hunger with such roots and fruits as they can find by search & occasionally a little wild honey. They carefully shun the society of civilized men, and will never approach a stranger except upon the offer of a piece of tobacco or a strip of cloth, for both of which commodities they show a great natural fondness. I have been informed however by a gentleman accustomed to visit the parts in which the Paliyans live, that a few of them have been induced to act as beaters for game & have to his knowledge considerably improved in personal appearance and intelligence since their employment and the receipt of their remuneration for their services. And such being case it would seem to be reasonable to hope that whole caste may some day be brought to adopt a more civilized mode of life. They are gentle in disposition and show no inclination to rob their neighbours.

It is always hazardous to speculate upon these matters, but it would seem to be no means improbable that these unfortunates are the descendants of some partially civilized caste which was centuries ago dispossessed of its hereditary lands on the hills by a merciless invasion or else nearly swept away by some ghasty epidemic, and being at once unable to acquire new lands on the hills and unwilling to descend into the plains rapidly sank into barbarism. There can be no question but that the Palani and Travancore mountains have been far more widely cultivated that they are at present: and many parts of India afford evidence of the fact that extensive tracts of country have some time been suddenly deserted by all their inhabitants, and from a state of careful, culture returned to one of nature. However, it may have ordinary Tamils in Physiognomy and physique that it is difficult to believe that they belong to another and earlier type.

According to census returns only two individuals of the caste existed in 1891 and this is of course not true. Several families are known to be existent, though how many it is impossible to guess". (Western Francis 1891).

***Analysis of the extract***

The following observations can be made about the Paliyan.

1. The Paliyans are not written as a tribe but mentioned as a mountain caste.
2. The Paliyan are considered much inferior or backward than the Poleiyans, which is considered as a Scheduled Caste.
3. They lacked the simple usages of ordinary men and without any house clothes nor any kind of property.
4. They were, in those days, leading a nomadic life searching roots, tubers, fruits and occasionally a little wild honey which was satisfying their hunger.
5. They carefully shun the contact of civilized people and approach a stranger only for a piece of tobacco or a strip of cloth. This indicates how primitive people were attracted towards civilized people.
6. Even in those early days they were serving as beaters in games. In this way even in those days a few of them got jobs as beaters and such people achieved considerable improvement in their personal appearance and intelligence with their employment and the receipt of remuneration for their services. The note also expresses the hope that the whole caste may some day be brought to adopt a more civilized mode of life.
7. In the second para, the note gives a hypotheses that the Paliyans might have been driven to this hill region either due to invasion or epidemic and they led to some extent civilized way of life. The hypothesis states that in case of Paliyans there had been secondary primitivization.
8. The Paliyans are like ordinary Tamils, in Physiognomy and Physique.
9. The Census report regarding number of Paliyans is ridiculous and shows the weakness of the enumeration of primitive people in those days.
10. Though the Madura manual has described the Paliyans as savages who decline to adopt the simple usages of ordinary men yet the note gives a sympathetic view of the Paliyans when it mentions that the Paliyans "are gentle in their disposition and show no inclination to rob their neighbours" (Madura Manual 1891). Some times these early writers are very disdainful towards Paliyans.

In a note on the Malai (Hill) Paliyans of the Madura District the Rev. J.E, Tracy writes as follows. "I went to their village at the foot of the Periyar hills and can testify to their being the most abject, hopeless and unpromising specimens of humanity that I have ever seen.........."Further Tracy writes, "if they were as ignorant as they were filthy...........The forest had been their home from time immemorial. Yet the forest seemed to have taught them nothing more than it might have been supposed to have taught the prowling jackal or laughing hyena. "After discussing this brief note of 1891 which is based on the Madura country of 1868 we can describe the general picture of the Paliyans as presented below.

## (b) Meaning & origin of the Paliyans

Etymologically the term 'Paliyan' may mean as inhabitants of Palni hills. It seems they were the earliest inhabitants in Palini hills or at least they settled in Palni hills for a fairly long period (Census 1961).

While narrating about the origin of Paliyans, Dahmen one of the early ethnographers on Paliyans go to the extent of stating that they cannot give any detailed information about their origin. But he narrates a folk tale of Tirunelveli in which Paliyan origin is linked to other communities such as Kaniyans, Malai Kuravars and Malai Valars. Similar folktales narrating the origin of Paliyans such as Mannadiar, Achari and Puliyans is reported (Mohanty: 1995) On the contrary Gardener (1972) comes out with two hypotheses i.e. "Perhaps these prople are nothing more than runaway Indians whose culture is secondarily simple due to exigencies of flight. "And his other alternative hypothesis is that these persisting enclaves of shy people possess a relict culture." While contradicting both his hypothesis due to the lacunae of empirical data he comes to a conclusion that Paliyans though they try to retreat into isolation still try their best to retain contact with their neighbours for economic reasons. Luiz in his brief ethnographical notes of Paliyan narrates. "They claim that their name originated from Valli a Pallichi (Paliyar female who was the wife of lord Subramanian, son of Siva) represented as Karthikeya of North India."

## (c) Language of the Paliyans

It is very difficult to guess the language that they spoke originally, as they have adopted now the Tamil language. But then at the beginning of the century they used to speak an unintelligible Tamil with "words so defaced and deformed that it takes one some time to be able to understand what they say".

Even today they speak Tamil among themselves which is unintelligible. They speak Tamil with a peculiar intonation which renders it scarcely intelligible. (Western Francies-1906). Further, Thurston writes about the language of the Paliyans as "They speak Tamil with a peculiar intonation which recalls to mind the Irulas" (Thurston 1909).

They are in the habit of signaling to each other in the jungle by different intonations of a cry much resembling 'COO-ee' in order to signify danger, pleasure etc.

### (d) Physical feature

While describing about the physical appearance, Dahman writes both men and women are on the whole markedly small and particularly the women are shorter. A little more than 5 feet to be the average height. Black bushy hair, snub nose and thickish lips are other features common in a more or less pronounced degree to nearly all Paliyans. The lips are thick and everted, eyes are dark, the average facial index is 84.9 while the average circumference of chest is 74.9 and average span of arms is 167.3 (L.a. Krishnan) Like other primitive tribes, the Paliyans are short of stature and dolicho cephalic and the archaic type of nose persists in some individuals.

### (e) Medicine, Health and Longivity :

Describing the causes of ailments Dahmen writes.

"While hunger may be assigned as one of the chief causes of their stunted growth and of most of the ailments of our Paliyans on the other hand intemperance in eating after long days of privation, often also the quality of their food; rotten fruits stagnant water and others. Damp and cold and in certain places, malaria may be added as not infrequent sources of ailment. Against ailments the Paliyans have their own remedies: infact, some Paliyans have made a name for themselves by their knowledge of medicinal properties of herbs and roots, for instance, they make a white powder known to be a very effective purgative from some varities of roots. Against snake bite they always carry with them certain leaves which they hold to be a very efficacious antidote yet, in spite of these and various other remedies, there are disorders before which the Paliyan is quite powerless. Such he attributes to wicked and malicious spirits who must be propitiated by sacrifice. Of all diseases small pox is the most dreaded by him. (Dahmen: 1908 :22).

In census of India 1931, it is stated, "it seems to be a general circumstance for Paliyans (other than Christians) to be peculiarly attached to their tribal life and customs and to be hardy and long-lived" Mr. Vedanayakan thinks that there must be more than one Paliyan centenarian.

### (f) Appearance and dress

The following extracts would indicate the dress and appearance of the Paliyans. "Their clothes are of the scantiest and dirtiest, and are some times eked out with grass or leaves (Western Francies 1906). Thurston writes. *"The men wore filthy* loin cloth only, but the women were fully clothed like the women of the plains, in very dirty saris. They were probably doled up for the occasion, as I have always seen them in the jungles wearing *nothing but a lion cloth."* Rev. Tracy writes "They were barely clothed, their rags being held about them, in one or two cases, with girdles of twisted grass——but they seemed to have had no better times to look back upon".

### (g) Ornaments

Paliyans love ornament to adore their body particularly during festival etc & they also make designs in their body out of white pigment, obtained from plants in parts of their body such as forehead, breast, arms, the men generally in only one of these places at a time, the women often on the three. (Dahmen: 1908:25). Besides tattooing Dahmen describes how Paliyan women decorate themselves with necklaces of shinning metal and glass-beads and adorn their hair with wild flowers or, with odoriferous leaves and lemon grass particularly during great festivities. In 1961 census it is mentioned:-

> There is also nothing distinct about the ornaments worn by the Paliyans. Women wear a few strings of glass beads and corals. They pierce the nose and ear – lobes and wear cheap stone-studded ornaments. A few wear koppu in the upper part of the ears. In the wrists, cheap bangles are worn. All women in Shenbagathope wear silver anklets. Generally, the Paliyan women can be distinguished more by the absence of jewels than by any peculiar characteristic of the jewels, worn by them.

## Economic life of the Paliyans

### *(a) The nomadic life and occupation*

All the earlier ethnographers have described the nomadic habit of the

Paliyan who roam round the hills and forest in search of tubers yams and in search of wild animals for hunting. Rev F. Dahmen writes.

> The Paliyans are a nomadic tribe who for the most part move in small parties through the jungle-clad gorges that fringe the upper Palani Plateau (6000 to 8000 feet high). There, they maintain themselves mostly on the chase and on roots, leaves and wild fruits, and also by hiring their labour to the Kunnuvan or Mannadi villagers to actual through relatively recent owners of Palni village lands. According to the Madura Gazetteer 1906 some bands of Paliyans are also found roaming in the fields of the Varshanad range, the line of low hills that face the Palnis and are, like them, the offshoot of the western ghats. The Tirunelvelli mountain too, it appears, possess some specimens of this interesting tribe. (Dahmen: 1908: 19-20) Hunting and honey collection are interesting occupations besides collection of yams, fruits etc.
>
> Thus for hunting and food gathering they adopt a nomadic life although they are confined to particular patches. Describing the settlement of Paliyans in the missionary estates of lower Paints Dahmen writes that, "Though so far reclaimed from the savage state, they have however, retained much of their former habits, customs and ways of living, above all their indolence and love of adventure and independence. Even now at times, as was but lately the case with one of our Christans they avail themselves of the first opportunity to slip back into the jungle and enjoy once more the free and hazardous life of the woods (Dahmen: 1908: 20) General Burton describes the Paliyans as good trackers and many of them carried bows and arrows and a few even possessed matchlocks." (Thurston: 1909).

The semi nomadic life of the Paliyans is mainly due to their economic pursuits of hunting and food gathering habits. Though at present they are undergoing rapid chages, the traditional occupation of hunting, food gathering, collection of minor forest products particularly that of collection of honey are their special occupation which they follow in different degree in different areas. In the chapter dealing with ecology and in the economic chapter their food gathering habits and associated tools, hunting and fishing methods, snaring and trapping of birds and animals, collection of honey and minor forest products will be discussed along with their semi nomadic habits. Again in the same, economic organization we would discuss how they are adopting new avocations along with their traditional occupations.

Regarding their tools and technology we get description by different writers towards the end of 19th century and early part of 20th century where they mention that their implements of hunting and gathering were very few and simple. Some emphasize that they had not even bow and arrow or other fishing implements yet others have referred how they use guns even stealthily. Penha writes "an axe, a knife and a pot are all the implements he carries," Of course they used to make rough and ready made digging sticks for digging roots and tubers as and when necessary.

### *(b) Material culture in the past (Dwellings)*

Dwellings of the Paliyans: Paliyans as hunters and food gatherers till the recent days were paying very little attention for construction of good houses. They mostly depended upon caves, trees, or on rough and ready made leafy huts. Their huts were built entirely of grass and consisted of one room each and that open at the ends (Thurston: 1909). The district gazetter of Madurai Vol-I contains a brief description of their dwelling, "they sometimes build themselves grass huts but often they live on platforms on trees, in caves or under rocks." Dahman writes: Paliyans in the woods stay in a cave, a dolmen, a platform on a tree or huts which has no particular model. Some of them consists of merely two inclined planes of sticks covered with grass or mats or resting on the ground: in others grass roof is supported on poles and the interstices between them is filled with twigs and boughs (Dahmen: 1908: 25).

Among the Paliyans in the Kerala side working in the estates, the houses are reported to be like Mannan pattern with dimension of 20 ft x12 ft and are made up of reeds, while the roof is thatched with leaves of reeds they have fire in one corner or of the hearth (L.A. Krishna Aiyar). In Thurston we have the description of the dwelling of the Paliyans as follows.

During the dry season they live in natural caves and crevices in rocks, but, if these leak during the rains they erect a rough shed with the floor raised on poles off the ground, and sloping grass roof, beneath which a fire is kept burning at night, not only for warmth, but also to keep off wild beasts. They are expert at making rapidly improved shelters at the base of hollow trees by cutting away the wood one side with billhook.

### *(c) Food*

Desirous to see how Paliyans procured their food Dahman accompanied with the Paliyans to the forest to witness how they collected their food. He writes "they collected roots of various sizes which were simply put to the fire & when they turn black it is taken out and eaten. This is practically their only preparation when living in the woods. Herbs, the leaves of certain trees, wild fruits in particular those of wild date tree (*Phoenix farinfera*) very common in these parts, supply them with food also"

He writes that Paliyans relish to take mushroom (ten varieties) apart from honey. R. Faulkes writes that the staple food consists of roots (the wild yam chiefly) honey and the flesh of animals and birds. They also eat various grains (cumbum, cholam etc) when they can get them either from minor forest produce contractors or by going down to the plains themselves. They exchange honey, roots, skins etc for what they require. Among the Paliyans in the Periyar state of Travancore apart from taking flesh of birds and animals (except bison). 'ragi' forms the staple diet and rice is taken only when it is offered by the cardamom estate in exchange of their labour (L.A. Krishna Ayer).

### *(d) Trade and Economy*

Markham's 19th century work narrates the silent barter for honey between plainsmen and mountain tribals of Madurai District. Even today overlooking the railways in the erstwhile district of Madurai (centre of ancient Tamil Sangam poetry and a trade partner of Augustaine Rome) we can still see dispersed bands of nomadic food collectors, the Paliyans.

In the earlier accounts of Paliyans most of the writers have emphasized that trade is limited and restricted to the level of barter of forest commodities with some bare necessities (Dahamen: 1908:25). For instance salt was obtained through barter or sometime purchased from itinerant hucksters. Paliyans in their free time engage themselves with the preparation of small bird cages or in weaving a rough kind of mat or in basket making. The small nicknacks they turnout are made according to rather ingenious patterns and partly coloured red and green vegetable dyes. These with the skins of animals and odoriferous resin collected from the Dammar tree are about the only article which they barter or sell to the inhabitants of the plains or to the Mannadis, the civilized villagers of the Palanies (Dahmen: 1908: 25).

### *(e) Major economic activities of the Paliyans*

Most of the accounts of economic activities of the Paliyans are rather sketchy and written negatively. Thus Penha writes, "The Palliyans cultivate nothing, not even sweet potatoes. He keeps no animals except a stray dog or two. An expert honey hunter, he will risk his neck climbing lofty precipices or precipitous cliffs".

Describing their economic life Dahmen narrates - "The Paliyans are nomadic tribe, who for the most part rove in small parties through the jungle clad gorges that fringe the upper Palnis plateau. There they maintain themselves mostly on the products of the chase and on roots (yams etc) leaves and wild fruits" (e.g. of the wild date tree), at times also hiring their labour to the Kannuvan or Mannadi villagers (Dahmen: 1908:19).

For the collection of minor forest product we get the description,

> They collect for the Forest Dept. minor forest produce in the form of root or root bark of Ventilago madraspatana and Anisochilus carnosus the fruit of Terminallia Chebuta (myrabolams), honey bee wax etc, which is handed over to a contractor in exchange for rice, tobacco, bettle leaves and nuts, chillies, tamarinds and salt. The food thus earned as wages is supplemented by yams, (tubers of Dioscorea) and root which are dug up with a digging stick, and forest fruits. (Dahman-1908).
>
> These early writers have also given detailed description of various method of hunting honey collection and fishing which would be discussed in second chapter dealing with ecology and in third chapter dealing with economics.

## Social life of the Paliyans

### *(a) Marriage*

W. Francis writes, "weddings are conducted without ceremonies, the understanding being that the men shall collect food and the women cook it". Referring to the Paliyans who live on the Travancore frontier near Shenkotta Penha writes. " At a marriage, the father taking the hand of the bride, and putting it into that of the bridegroom, says "I give this girl to you. Give her roots and leaves and protect her". The value of bride or bridegroom, depends very much on the quantity of roots, etc. which he or she can collect. When a widow does not remarry, the males of the community supply her with roots and other products of the jungle. Marriages are as a rule, contracted within the settlement, and complications occasionally occur owing to the absence of a girl of a suitable age for a youngman. Indeed in one settlement, I came across two brothers, who had for this reason resorted to the adelphous form of polyandry……Paliyans admit members of other castes into their ranks." Marriage takes place soon after puberty and never before. Marriages are permitted between a man and his maternal uncle's daughter or sister's daughter. (Thurston: 1909). Dahman writes," The laws of marriage are so loose that the true marriage can hardly be said to exist. Husband and wife separate as freely as they join. The children of their previous marriage remain with their respective fathers".

### *(b) Family*

"As to the relations that exist between man and wife, they also still savour much of the primitive times. Women are considered little better than animals, and rather harshly treated and in some instances brutally beaten" (Dahmen :1908: 27).

### *(c) Life cycle rituals*

(i) **Birth rituals:** "There seem to be no birth ceremonies besides purification. There are no midwives. The mother or some other elderly woman, does all that is necessary at a confinement. The new born baby is plunged into the nearest stream for its first bath and the mother also bathes immediately after her confinement and washes her cloth. For some days, after delivery she and the baby sleep on a bed of ashes, probably for warmth. Oil is never administered to the baby as purgative but a decoction of herbs known to Paliyans is given when necessary. Other decoctions are prepared for other baby illness. "The child is simply named after the place where or near which, he is born……… To these children the parents show great affection nursing and feeding them with great care and taking a short of pride in keeping them fat and pump". (Dahmen 1908; 27).

(ii) **Puberty:** When a girl attains maturity, a grass hut is built for her. There she is shut up for 12 days, food being brought to her once or twice a day. In the morning of thirteenth day the matrons of the settlement forcibly drag her to some neighboring pond or stream and plunge her into the water. The girl is once again brought back to her hut and confined there till two days, during which time no food is given her. The fifteenth day she is at last set free; the hut where she lived is burnt down and a grand feast takes place (Dahman: 1908:27).

**(iii) Death rituals:** The dead are always buried. The body is not washed or prepared in any way and is kept in the hut or cave where death occurred until the time of burial. The body is placed in the grave, which is only breast deep, in a recumbent position on its back. When a death occurs messengers are sent to neighboring habitations to announce it. All are bound to attend it, especially the near relations. (Faulkes).

Francies (1906) writes" when one of them dies the rest leave the body as it is and avoid the spot for some months". The body in the meantime is left unburied even though it becomes offensive. The dead body is borne to the grave by any one while the front bearers are the near relations. A penalty is inflicted on those who do not attend the funeral.

On the eighth day after the death, the spirit of the departed is worshipped with offerings of roots and honey and its protection is invoked. From that time onwards the spirit is deemed to be merged in the deity.

## Social control and leadership pattern

**Leadership:** While visiting a Paliyan village at the foot of the Periyar hills Rev J.E.

Tracy writes:

> The chief man of the community was an old man with white hair. His distinctive privilege that he was allowed to sleep between two fires at night while no one else was allowed to have, but one a distinction that they were very complaisant about, perhaps because with the distinction was the accompanying obligation to see that the community fire never went out. As he was also the only man in the community who was allowed to have two wives (Thurston: 1909).

The Paliyan society is fairly liberal in case there is violation of social norms or life cycle rituals. Regarding the leadership pattern in the social organization of the Paliyans Dahmen informs "at the head of the different groups there is a chief called Nattanmaikaran i.e Chief of the country. Perhaps, this institution was borrowed from certain other castes of the plains. Nattanmaikaran is greatly respected and annually offered presents such as honey, skin or piece of cloths etc. His dignity is hereditary. Besides, the chief the other official is called Tandalkaran i.e. the Collector of the tribute, who comes regularly in the name of the Chief to receive the contribution of Paliyans and at times to preside at their social and religious feasts and ceremonies.

Thus the major activities of Paliyans leader were to maintain the norms of social practices particularly in life cycle rituals, such as marriage and death rituals. By such roles Natanmaikaran integrates the village community.

## Role of leaders in life cycle rituals

### *(a) Marriage*

As regards marriage Thurston writes

As a rule marriages are arranged for the; young people by the elders of a family. But there are cases when the young couple make their own choice, generally as the result of illicit sexual intercourse. Such behaviour is condoned without much difficulty and a marriage is brought about as the best solution of the trouble.... Widows may remarry, Adultery is not common, but when it does occur, no penalty seems exacted, although the act is strongly condemned...

Marriages are permitted between a man and his maternal uncle's daughter. All other marriages among relations are prohibited. An alliance between interdicted relations is viewed with displeasure and is treated as concubinage, but is condoned when due apology has been made by the offending couple. (*Thurston: 1909*).

Paliyans seem to be very strict as regards social obligations such as death rites although the punishment is waived when satisfactory explanation is given.

### *(b) Death rites*

A penalty is inflicted on those who do not attend the funeral. The delinquent is made to kneel, a heavy stone is placed on the nape of the neck, and he is asked to explain why he was absent. This is generally satisfactory. The offender and the chief mourner exchange a drink of water from each other's hands and the affront is thereby condoned".

## Inter-Caste relationship of the Paliyans

There are many documented references in which a caste man marrying a Paliyans woman is narrated. Penha writing on the custom of the Paliyans who live on the Travancore frontier narrates, "The Paliyan is just a shade lower than the Eluvan. He is permitted to enter the houses of Eluvans, Eluvanians (betel growers) and even of Maravars, and in the hills, where the rigor of social code is relaxed to suit circumstances, the higher castes even drink water given by Palliyans and eat roots cooked by them".

In the Madura Manual, it is noted that "the Paliyans have always been the praedial slaves of the Kunuvans." Though they depend on the chase roots leaves and fruits of the forest, at times also by hiring their labour to the Kannuvas or Manadi villagers. Thus they had close contact with higher castes from long past. By abdicating beef eating they could raise their ritual status and therefore could mix with higher castes more freely, in their social life.

## Religion of the Paliyans

**(a) Religion and Magic:** The primitive religion is designated as animism as coined by Tylor. Subsequently as per different, situations, terms like manaism, shamanism etc. were also coined to describe primitive religion. As per Indian situation, the term religions practices of little tradition as described by Redfield in his peasant studies, becomes more popular to describe religious practices of both tribal and rural societies. However, in the earlier phases of colonial rule the theoretical sophistication of later stages had not emerged, yet such descriptions were based on direct observation and gives us a picture of traditional religious practices of a tribe and its subsequent changes, from scattered records of early administrators and Christian Missionaries. The Paliyan religious practices are presented below with an effort to organize it in a systematic manner.

**(b) The Paliyan pantheon**

(i) **Pallichi Ammal:** The Paliyans worship Pallichi Ammal. They have no knowledge of any other god or goddess. At rate intervals the whole community worships together this goddess, but each family worships her separately more often. An old man said communal worships occurred once in 10 to 20 years, but others said it was more frequent.

**Rituals:** A few stones are set up in a row, the number being immaterial though generally from one to seven, under a rock or tree. The stones are usually smooth pebbles from a stream. Each stone is painted with a figure to represent a man (or a woman) with vermilion paste. The worship consists of the offering of honey and some roots of the wild yam. Then they prostrate themselves on the ground before the stones and utter prayers for protection from wild beasts and evil spirits. The women are also present and prostrate.

**(ii) Mayandi:** Describing the religion of the Paliyans Dahman observes. The principal of these religious ceremonies takes place about the beginning of March. Tandalkaran (collectors of the tributes) gives notice before hand to prepare for the sacrifice to the god. The god is called Mayandi. Usually it is represented by a stone, preferably one to which nature has given some curious shape, the serpent form being especially valued. 'The stone that represents him has its shrine at the foot of a tree, or is simply sheltered by a small thatched covering.

**Ritual:** On the appointed day, the Paliyans gather before sunrise, fire is made in a hole in front of the sacred stone, a fine cock brought in decapitated amidst the music of horn and drum, and the blood made to drip on the fire. Though they profess to worship only the god Mayandi, their belief in, and dread of evil spirits, their magical practices and superstitions show their religion to be rather demonical in character and spirit.

**(iii) Palini Andavar:** The other important religious practice is to visit the famous shrine of god Subramaniam, Siva's son situated in the Palni-Hills.

They go in small bands along the rugged mountain paths, each carrying a pair of new sandals on his head, march to the holy city with the monotonous sound of horns and drums. Before handing over their offerings of sandals to be suspended in the shrine they piously strike their own heads with it.

**(iv) Karupuswamy:** G.F.D. Penha writes that the Paliyans regard sylvan deities with great veneration Karupuswamy is the tribe's tutelary god. When a great haul of wild honey is made, offerings are given at some shrine.

**(c) Nature of Private or individual worship:** The Paliyans apart from the congregational form of worship also pray privately to Palichiammal for protection against danger of any kind in the jungle. This is done mentally and without any outward ceremony. Although they evidently dread evil spirits, their ideas about them are very vague and they can not describe them. They have no permanent temples, and the locality where the community worships is chosen according to convenience and need not always be the same, community worship seems to take place when the community can afford it and such occasions are, therefore, rare.

**(d) Shamanism:** R. Faulkes describes the shamanistic practice of the Paliyans of Madura without referring it directly. After performing their worship before few pebbles they pray for the protection from wild beasts and evil spirits. The chief performer is generally an elderly man who is called the "dancer of the god" is or 'goddess' in Tamil. He works himself upto a state of ecstasy and falls into a trance. The 'Sattan' another elderly man, then acting as interpreter, puts himself questions about the future welfare of the community, or of an individual and the "Thevaradi", who is supposed to be inspired by the goddess answers. *(Faulkes. R. 1932).*

**(e) Magic and superstition:** Dahmen writes that the powers of witchcraft and magic are attributed to the Paliyans by other castes and probably believed in by themselves. Dahmen also gives an illustration of their magical device adopted by them to protect themselves from the attack of wild animals, particularly that of panthers.

> Four jackals tails are planted in four different spots, chosen so as to include the area within which they wish to be safe from the claws of the brute. This is deemed protection enough: though panthers should enter the magic square, they could do the Paliyans no harm; their mouths are locked. (*Dahmen 1908:30*).

## Acculturation and Sanskritization

Though Paliyans are depicted mostly as hunters and foodgathers during the colonial period, we do not have any reliable information about their customary life style and avocations during the pre-colonial days. However through the analysis of their customs and traditions of colonial period it appears that they had long contact with the higher Hindu culture and traditions from time immemorial, right from the Sangam literature and from the historical evidences of Roman contact and more particularly description of silent trade through barter, it appears that the tribal communities of Tamilnadu had access to or contact with more civilized communities in those earlier days. But most significant was the acculturation process with the great Hindu civilization which are also referred to in accounts documented during colonial rule. Specially the interaction with the great tradition of Hindu religion as noted above is more spectacular and the process of Sanskritization was initiated and accelerated in the Paliyan culture. Thus the Pallichi Ammal of the Paliyan was perhaps sanskritised as Valli and became the consort of the god Subramaniya. "The Tinnevelli Paliyans say that Valli, the wife of the god Subramaniya was a Paliyan woman. *(Luiz: 1962).*

They also worship Siva. "They pretend to be the followers of Siva, and always attend the Adi Amavasai ceremonies at Courtallum".

From very early days they had modified or reformed their food habits. They do not eat beef and consequently carry no pollution (Western Francies). As they carry no pollution, they are some times employed in return for food, as night watchmen at the Vaishnavite temple known as Azhagar Koil at the base of the hills.

Thus the deity like Karupuswamy might be some tribal god which is adopted in the local folk cultures of the Hindus. The high gods of Hindu religion with all India spread had been adopted and worshipped or respected by the Paliyans and thus could attain ritual purity and could achieve higher status even in caste Hindu society.

It seems Paliyans came under sanskritization at the very early stage. We can make a hypothetical reconstruction of the past history of the Paliyans based on current folk tales and other usages. We can discuss about it at a later stage. Now, we would focus how the description of these early writers were focusing the Paliyans as a tribal community through their exotic practices and primitive custom.

## SECTION – III

## ANALYSIS OF TRADITIONAL PALIYAN CULTURE

The colonial period during which ethnographic description relating to Paliyans are recorded is not a brief period and thus tremendous change has occurred to the community. As mentioned earlier the Paliyans in the first "manual" of the District of Madura, the Madura country (1868) and in the subsequent Madura Manual (1891) the Paliyans are not written as a tribe but simply mentioned as a mountain caste. The significant change in designating Paliyan as a tribe becomes conspicuous from the beginning of the 20$^{th}$ century.

It seems that the early administrators were influenced by the local custom of designating the Paliyans as a caste. Even in Tamilnadu the word 'jathi' is frequently used, as in other parts of India, for mountain people and forest dwellers.

Thus the use of the term (tribe) in place of 'jathi' is a new innovation during the colonial period which brought tremendous change in the approach and understanding of the community by the common man and the administrators. In the following pages we would try to explain how the concept tribe was introduced into Indian society by the colonial rulers.

## Application of the concept 'tribe' against the Paliyans during the colonial period

The concept 'tribe', as it is understood at present, was not there in pre-British era. In those days, every community whether big or small, whether nomadic or settled, whether advanced or backward were categorized only as 'jati' or caste. But subsequently the concept 'tribe' was introduced by colonial rulers to most of the communities who lived more or less in hills or forests. Subsequently due to political exigencies, the empire builders tried to develop a tribal policy and tried to emphasize how tribal society is somewhat different from the main stream of Indian society.

By analyzing the documents of the colonial period an effort is being made to identify the features by which the Paliyans were identified as a tribe, though there is no mention of such criteria directly, yet, we can identify these features through analysis of Paliyan habitats, occupation and culture etc. Through the analysis of such ethnographic accounts of the colonial period, we can identify three features of Paliyan society which can be taken as criteria for identification of Paliyan as a tribe.

Fruther in subsequent stage, an effort will be made to go back beyond the colonial period in order to explore what might have been the situation of the Paliyans during the Pre-British era. Of course almost no document is available relating to the pre-colonial era yet with the help of ethno-historic method and analysis of folklores a hypothetical reconstruction of their origin and historical development is attempted below.

In the final section, the changes that occurred in the Paliyan society, their problems and their adjustment with the changing situations would be discussed.

Along with the analysis of socio-economic changes during the colonial period, problems of the Paliyans would be discussed at the end.

As mentioned earlier the three features of Paliyan culture and society which probably accounted for the categorization of the Paliyans as a tribe or primitive community are mentioned below:-

(1) The environmental features of the Paliyan habitat, their isolation and relative lack of communication with plains people, and their nomadic life alongwith food gathering economic segregated them from more advanced settled communities.

(2) Further the exotic and archaic features of their culture were also instrumental in identifying the Paliyans as a tribe.

(3) The negative and disdainful attitude of the early administrators and ethnographers (perhaps also shared by the surrounding higher caste local communities) to some extent led to their categorization as a primitive community.

These three features would be discussed below briefly from the profile of Paliyan culture as presented earlier.

**(i) Isolated habitat and nomadic life:** In a colourful introduction to the Paliyan habitat & culture Dahmen writes,

> to the ethnologist, the Palni Range, like other ranges of southern and central India, offers interest of another sort. It is one of those last mountain homes where the fast disappearing remnants of primitive races have taken refuge before the sweeping tide of more civilized invaders. Within these fastnesses, as "in a vast museum of nations" these scanty relics of a bygone world are yet to be found with all their primitive customs and religions, still safe and hardly affected by contact with their neighbuours *(Dahmen;* 1908:19).

Emphasizing their isolation Dahmen writes. "The Paliyans are a nomadic tribe, who for the most part rove in small parties through the jungle clad gorges that fringe the upper Palnis plateau"

On most of the occasions the habitats of the Paliyans are either at the foots of the Western Ghats or on high plateaus near the forest areas. Since they were basically food gatherers in those days they were leading almost a nomadic life digging roots and tubers or collecting fruits and herbs from the forest, collecting honey & hunting small games. Because of such life style they were considered backward & were categoriesed as a tribe.

**Comments** - Though the Paliyans live in or near the forests they had close relationship with agricultural castes who migrated to hills of Western Ghates some four or five hundred years back. They had reciprocal relationships with these caste Hindus.

**(ii) Archaic and exotic features of the Paliyan culture:** The early ethnographers and administrators were fascinated by some of the unusual and strange practices and archaic customs of the Paliyans. Description of all such customs and practices were given prime importance in those days. As the intellectual atmosphere of the period were influenced by the prevailing evolutionary ideas and filled the ethnographers with romantic feelings, such primitive customs of the Paliyans were focused by the early writers. They tried to emphasize the special characteristic features of their community in almost all aspects of their culture starting from their material culture to their socioeconomic and religious life. Yet such descriptions also contained valuable informations. Hence the traditional culture of the Paliyans is described based on such documents.

(a) Material culture of the Paliyans: Though quoted earlier, material culture may be repeated for clarity. During earlier days the Paliyans were not constructing any durable houses permanently. They were living either in the caves or in a hut under rocks or dolmen or in some temporary rough and readymade houses consisting of two planes of sticks resting on the ground covered with grass or mats.

As regards clothing it is mentioned that males used to cover their loin with a bunch of grass or leaves or with some rags while women used to wear a sufficiently broad band of plaited grass or palm leaves. And about cooking it is said that they used to dig a pit and put roots into it while a heap of wood is burnt over it throughout the night. Their method of hunting fishing & honey collection are equally interesting & ingenious which would be presented in the 2nd & 3rd chapters.

(b) Social life of the Paliyans: As regards marriage Dahman writes, "The laws of marriage are so loose that true marriage can hardly said to exist. Husband & wife separate as freely as they join".

> "As to the relations that exist between man & wife, they also still savour much of the primitive times. Women are considered little better than animals, and are rather harshly treated and in some instances brutally beaten." *(Dahmen: 1908:27).*

The gazetteer writers also try to search for unusual situations of some stray cases and speculated the existence of adelphic polyandry. Though marriage rituals are simple, the death rituals are quite elaborate and these are strictly adhered to under the

> supervision of the chief mourner. Those relations who fail to attend the funeral are punished. The religion and superstition have been exhaustively quoted earlier. The rituals indicate their archaic features while the superstitions and magic to some extent indicate their backwardness and lack of refined thinking.

**Comments:** The material culture and social institutions of the Paliyans may appear archaic but such institutions are not very different from those of caste Hindus since most of such features are due to their adaptations to their ecology. There is nothing very unusual or exotic.

**(iii) Negative and disdainful attitude towards the Paliyans:** Besides giving emphasis to the archaic and exotic features of Paliyan culture, these early writers have, some times taken a disdainful attitude and criticized the Paliyan customs & manners. A few cases of such unsympathetic and scornful attitude may be referred below. Referring to the Paliyans Rev. Tracy writes".... And can testify to their being the most abject, hopeless and unpromising specimens of humanity that I have ever seen"

Pointing to their lack of cleanliness further he writes -

> A stream of pure water was flowing within a few feet of their huts & yet they are as fowl and filthy in their personal appearance as if they were more animals and very unclean ones.

Further describing their filthy habits he continues -

> They were as ignorant as filthy. They had place of worship, but seem to agree that the demons of the forest around them were the only being that they had to fear. They were barely clothed, their rags being held about them, in one or two cases with girdles of twisted grass.

Rev. Tracy writes that the forest had been their home from time immemorial. Yet the forest seemed to have taught them nothing more than it might have been supposed to have taught the prowling or the laughing hyaena. There were no domesticated animals about their place: Strange to say, not even a pareah dog. They appeared to have no idea about hunting any more than they had of agriculture. Besides he mentions that " they appeared to have no idea of beauty or solemnity of the place they had selected as their village site" . Such disdainful remarks show that early ethnographers were ethnocentric and lacked the sympathetic feelings towards these unfortunate people.

**Comments:** Such unsympathetic, disdainful writings show that the early ethnographers had poor impression about the Paliyans and they never tried to explore how in a state of nature they tried to adjust and adapt themselves. Such biased views to some extent created a cleavage between tribes and castes both of whom lived in same environment.

## Hypothesis regarding the Paliyans

**(i) Hypothesis regarding the origin of Paliyans:** As mentioned, the name Paliyan signifies that they were the inhabitants of Palni area. This indirectly signify that earlier this might have been a branch of a bigger tribe whose name they must have used before settling in Palani hills in course of their sojourn and settlement in Palni, they might have forgotten their earlier habitat and adopted the present Paliyan term. We can explore what might have been the name of the earlier stock from which the present Paliyans might have separated. We can discuss the possible tribes of Tamilnadu in order to identify the ancestral stock from which the Paliyans might have separated.

Malayali community is the biggest tribe in Tamil Nadu and they were fairly distributed in northern part of Tamil Nadu. But their folk tales indicate that Mudaliars of Kanchipuram beheaded the Vedan's and married their women & thus the Malayali tribe emerged. Whether such story is true or not the fact remains that the Vedans which is now a backward agricultural caste in Tamil Nadu might have been the original tribe which assimilated with the plains people & are designated as Malayali.

The nomadic pastoral community, 'Kurumba' are the other two widely distributed tribes in Tamil Nadu but they have no other cultural traits which can be compared with the Paliyan. A hypothesis can be made that they (Paliyans) were a section of Irulas of Tamil Nadu which is very widely distributed in the state. Besides they have some significant similar cultural traits with the Paliyans.

There are two streams of migration of Irulas in Tamil Nadu; the Irulas of western districts and that of eastern districts. Researchers think that both the streams are not identical although they might have been designated by outsiders by the same term.

The hypothesis suggested here is that, Paliyans had been segregated from the main stream, the western Irulas, probably some where around the Palani hills. Thus, the tribe Irula is in the North Western part of Tamil Nadu while Paliyan are distributed in South western part only.

The hypothesis that the Paliyans segregated from their original larger community, the Irulas, is based on the following facts.

(a) Both Irulas and the Paliyans seem to belong to Predravidian tribal communities.

(b) As per the ethnographers of colonial period the language of the Paliyans, atleast their intonations have similarities to that of irulas. As quoted earlier, Thurston writes about the language of the Paliyans "They speak Tamil with a particular intonation which recalls to mind Irulas" (Thurston 1909).

(c) (i) Both the Paliyans and Irulas are highly religious communities.

(ii) In Kerala, a tribe called Malpandaram is considered same as paliyans. Etymological meaning of the term Malpandaram is the priest of the hilly area. It indicates that at one stage Paliyans acted as priests. Irulas also acted as priest in Coimbatore district (Alyappan: 1948:102).

(iii) In Coimbatore district Padmabati an Irula girl, through her devotion became the spouse of lord Vishnu (a local folk tale).

It is found that in the Subramaniyans temple of Palani hill, the deity has a tribal spouse called Valli. In the Palani hill the lord Subrananiyam has two wives Devajani and Vedavalli or simply as Valli. Tirunelveli Paliyans say that Valli, wife of god Subramaniya, was a Paliyan woman. As per Dahmen, Paliyans are closely associated with Palaniswamy or the lord Subramaniyam of Palani. Thus the hypothesis is that a branch of Irulas at one point, in course of their expansion or migration concentrated at & around Palani hills and became a dominant community in Palani area. Later while they moved further to south or south west they retained their Paliyan title both in Tami Nadu & Kerala while a branch of them further migrated to east Kerala & became Mal Pandaram.

The above hypothesis regarding the origin of the Paliyans also serves to clarify the confusion between the three tribes Viz, the Paliyan, the Poliyan and Mala Pandraram. The ethnographers of 19th century (Ref Madura Manual 1891) consider Poliyans much more advanced than the Paliyans. But as per local traditions the Paliyans have much higher ritual status than the Poliyan since the later consumes beef without any inhibition, while the Paliyans strictly abdicate the consumption of beef on any occasion. Further the Paliyans are considered ritually pure and are used as porter by caste Hindus while visiting the pilgrim centres or temples situated on high altitudes.

The Paliyans – They enjoy high ritual status. Basically they are seminomadic or nomadic food-gatherers. They also work as herders or agricultural labourers. Their main source of income is through sales of minor forest produce.

Mal-Pandaram – Inhabit the hills or plateaus. Main occupations are collection and sale of minor forest produce, shifting cultivation and small scale or marginal chiefly eke out their living through agrestic labour. Both the Paliyans and Mal Pandarams are considered identical or closer to each other yet Mal Pandaram consider themselves much higher.

**Comments** - The above process of absorption of the Paliyans within the caste societies through the process of (1) migration (2) fragmentation (3) integration in to the caste societies through reciprocal economic relationship within a particular eco-system are interesting features in understanding the caste tribe interaction. But the colonial rulers were more interested to create a cleavage between tribals and castes and so could not notice the close interaction between castes and tribes.

**(ii) Other hypotheses regarding the Paliyans in Madura Manual 1891:** As quoted earlier in the Madura manual 1891 an interesting and important observation has been made.

It indicates the early speculation of an administrator who believed or had following observations.

(a) Palani hills was much more cultivated centruries ago.

(b) The present Paliyans are the descendants of partially civilized caste who were dispossessed of their hereditary land.

(c) Unable to acquire new lands on the hills rapidly sank into barbarism.

(d) As evidenced in other parts of India, the Paliyans returned to one of nature from state of careful culture.

(e) The Paliyans are like an ordinary Tamils in physiognomy & physique.

The above observations are really interesting & indicate that centuries ago the Paliyans lived in an extensively cultivated areas of Palani and were leading to some extent civilized life. They had long contact with their Hindu neighbours. In the long process, the Paliyans have adopted certain deities of great tradition & their rituals while the Hindus have also adopted certain Paliyan deities and rituals. The Paliyans stopped consumption of beef and there by they improved their caste status in the surrounding locality.

## Acculturation during colonial period

The early ethnographers were more keen to depict Paliyans as a primitive community and therefore they were not very keen to depict the process of change which transformed the Paliyans to a great extent. Yet in their writings there are indirect reference of the problems and the process of change that occurred in the Paliyan society.

Due to better communication and the penetration of administrative machinery in inaccessible areas of the Paliyan habitat, large scale population movement occurred in the homeland of the Paliyans. Further such inaccessible areas were open to outsiders and business communities infiltrated to such areas and introduced new monetized market economy. Such situations affected and transformed Paliyan culture and society. Finally due to missionary activities large scale Christianization process started both in the Palani hills and in the Tirunelvelli districts. Thus in these areas, there are tremendous transformation of Paliyan society and culture. Further, due to new forest laws, reservation of forests and establishment of coffee plantations, fruit orchards, and other agricultural estates, exploitation of forest resources became restricted and Paliyans had no other means except selling their labour. Previously a few of them hired their labour to Kunuvan and the Mannadi Villagers. But during colonial period Naidus, Nadars and Muslims etc. migrated to Palni hills in large number. In this situation Paliyans were forced to become labourer and occupied lower status.

Previously the Paliyans could exploit the forest and could dispose these forest products as they liked. But during the colonial period all minor forest products were to be handed over to forest contractors in exchange for rice, tobacco, betel leaves, nuts, chillies, tamarind and salt. In this manner the forest contractors could exploit the simple Paliyans and could fully keep them under their clutches.

Thus during the colonial period the Paliyans lost their freedom and their life was controlled, by forest contractors, forest officers, and other Government officials, business men and other well-to-do people of their locality, belonging to higher castes. Thus, it seems, from an earlier higher status they were relegated to a much lower status of wage earning labourer.

## Summary

The profile of the traditional Paliyan culture is based on documents written during the colonial period. It is divided into three sections.

In the section-I it provides the general introduction is which divided into three parts. In part (a) the problem of the study, and the objectives of studying or exploring the traditional Paliyan culture is presented. In part (b) the distribution of tribal communities in India and their broad divisions are presented while in part (c) distribution of Paliyans is presented. In section I(a), the following three objectives of studying traditional Paliyan culture are presented.

(1) By exploring the traditional Paliyan culture we intend to identify the original Paliyan culture uncontaminated by the subsequent large scale acculturation.

By identifying the traditional culture of the Paliyans we hope to identify such features which were considered as primitive even in those early days.

In recent days much emphasis is given to the study of Hunters and Gatherers. By identifying the traditional culture of the Paliyans we can compare and focus the special features of Indian Hunters and Gatherers who are enclosed within the Indian civilization.

(2) To analyse the speed and direction of social change amongst the Paliyans we hope the profile of traditional culture will provide the base line for such analysis.

(3) In the present developmental process we hope that the knowledge of traditional culture of Paliyans, can be used in identifying the felt needs and in accelerating the developmental process.

The second section provides the profile of the traditional Paliyan culture. In section – II (a) a general account of the Paliyans with a brief extract from the Madura manual and its analysis is given. Further it provides details of origin of the Paliyans, distribution, language, physical features, Economic and Social life of the Paliyans, Life cycle rituals, Social control and leadership, Inter caste relationship of the Paliyans, Religions of the Paliyans, Acculuration and Sanskritization.

The third section presents a general analysis and a few hypotheses. It raises a question why the concept 'Jati" which was used earlier was substituted by "Tribe' and what were the implications of such substitution. This section provides a hypothesis regarding the origin of the Paliyans. Basing on Madurai manual it also provides hypothesis that Palani hills were cultivated much more in distant past. The same manual also provides another hypothesis that the Paliyans are descendants of partially civilized caste who were dispossessed of their hereditary land.

Finally the analysis concentrates on the present acculturation process as a consequence of which Paliyans are gradually losing their freedom, economic security and social status. It discusses how from a free and independent life of hunter and food gatherers they have become labourers, some times bonded labourers and are being exploited heavily through monetized market economy.

# 2

# Ecology and Social Structure

## SECTION – I

## INTRODUCTION

Anthropologists, in their study of ecology, try to show the relationship between natural environment and the culture of particular ethnic group. Though they have undergone through various stages of theoretical approach, such as environmental determinism and possibilism they rejected both for lack of sufficient empirical evidence. A more scientific base for the ecological approach in anthropology was established by the contribution of Julian Steward's concept of cultural ecology. Netting in his Encyclopaedia of Anthropology further simplifies Stewards concept of cultural ecology.

> Cultural ecology considers the systematic interaction of factors of environment, technology and human behaviour using the methods of social science to analyze the process of adaptation and change in specific society under particular environmental condition. (*Netting R. M. 1996*).

Subsequently Steward's cultural ecology concept has been revised and expanded to incorporate more concepts from biological ecology and scientific methodologics. In this chapter the physical environment of the Paliyan habitat and how the Paliyans exploit their environment through their technology and how this subsistence pattern influences the social structure and other aspects of their culture, has been discussed.

### Environment of the Paliyan Habitat

The Paliyans are distributed in the Western Ghats of Tamil Nadu excepting a few who migrated to Kerala. The tribe is distributed in the hills of Coimbatore, Anna Dindigul, Madurai, Tirunelvelli and Kamarajar district in Tamil Nadu and occupy hills which run from 8' 15' to 10'30' North latitude and from 76'50' to 78'5' East longitude forming an area roughly 150 by 60 miles (Gardener).

Western Ghats is considered to be the second largest mountain chain in India. The distribution of the Paliyans in this region can be observed in two areas.

(a) The massive Palani hills constitute the largest hill range in South India and along with Anna malai it spreads from east to west. The Palani hills rise to 8300 ft with several peaks of 8000 ft. and plateau of 105 sq miles at 7500 feet. This zone is the home of half of the Paliyan's population. (Gardner).

(b) From this massive east west range starts two ranges of Cardamom hill and Western Ghat hill ranges. The cardamom hill range is rarely used by the Paliyans while the parallel hill range, which situated to its east, spreads southward with peaks of 6000 ft gradually diminishing to 3000-4000 ft. and then continues almost without peaks. The smaller broken ranges spread towards south for about 160 kilometres until the extreme tip of India is reached. In the quiet lower valleys of the eastern slope the other half of the Paliyan population make their home.

Rocks of Western Ghats are among the oldest rock formation in the world. Geologists describe these hills as "large massifs of granitic rock of which Charnokites predominates. Mineral wealth is slight except magnesite, Chromite and iron ore (Gardner 1965-5 quoted from Krishnan 1960 iii, Pichamuthu 1953-45). Therefore plains people are not interested to penetrate to these hills for the removal of mineral wealth. The upper altitude is however, a potential attraction. As per Baliga,

> The soil in the top of Kodaikanal and Upper Palanis is usually a layer of black peaty earth over yellow clay. But in the lower hills it is generally shallow, stony and dry....(*Baliga: 1960:8*).

The plains people are usually attracted to these high hills for its fertile soils at the high altitudes while the Paliyans live in the stony dry lower hill slopes.

In the Palani hill alone, there are thirty large perennial streams. Besides there are large number of purely seasonal streams that swells in south-west monsoon but dries up in summer, creating great hardship to the Paliyans. The hill enjoy two seasonal monsoon rains that of south west and retreating north-east monsoon while the eastern slopes of Western Ghats are mostly in the rain shadow area with scanty rainfall. Since

Paliyans live in valleys at the lower altitude they suffer maximum water scarcity during the summer. But the position of the hills is the most important factor as regards availability of rainfall.

The Paliyans live closest to the forest environment and make optimum use of biological resources. Their dependence on plants is higher in comparison to their dependence on meat or fish. Since the forest is most important in the Paliyans life the vegetation profile of southern hills of Paliyan region is presented here.

Puri (1962) has presented six vegetation zones of southern hills. Gardener has tried to adopt it and further correlated this with Paliyan habitat (Refer Chart in Appendix-1). Paliyans very often live in relatively dry zone situated at the low altitude or at the foot of the hills, since it helps them to exploit the forest resources with convenience. Brief description of the zones is given below.

**Zone-I:** (1000-2000 ft). The dry thorny forest occurs in rain shadow areas. These areas have lateritic and basaltic granitic rocks with poor soil. The forest is mostly covered with species of Acacia with an open growth. Due to over grazing, most of the plants and trees belong to thorny variety only with few climbers and grass in rainy and winter seasons. Further with cutting of woods the situation is aggravated. There are several varieties of Dioscoria yams, which the Paliyans use as their staple food. The soil is very hard and stony. Therefore Paliyans have to undergo great hardship. In this zone also the Paliyans find trees in riverbank or moisturized area out of which they make their digging stick. Many Paliyans live in this zone.

**Zone-II:** (1000-3500 ft). It contains dry tropical evergreen forest more dense than zone-I. Yet it is also dry, stony and thorny. The trees are of 30'-40' height with thick leaves that form a canopy. Yet forest is a mixed one with a number of spiny and shrubby species. Due to constant grazing and browsing there is also the presence of thorny forest. This type of forest contains large number of valuable and useful trees. The Paliyans live in this Zone of forest mostly where wild yams and Rubiaceae trees are abundantly present, out of which they make digging sticks. But due to the illegal exploitation by woodcutters and poachers the valuable trees like rose wood and sandal wood are decreasing and faunal wealth is diminishing at present.

**Zone-III:** Moist Tropical forest: It has sporadic distribution depending upon rainfall. From the forest of this zone Paliyans exploit the bamboo tree, silk cotton tree, wild jackfruit and mango trees. Dioscorea yams are also found in this area.

**Zone-IV:** (3000-5000 ft): It is fairly wet but more temperate sub-tropical forest. The Tamils exploit the rich forest of this area from the nearby plain area. Yet in recent year's considerable number of Paliyans have moved upto coffee and orange plantation areas for salable forest produce and herding but rarely they inhabit in such areas in spite of the presence of wild yams and digging stick woods. The main tribal inhabitants of this zone are shifting agriculturists such as 'Muduvan', 'Malasar' and 'Mannar'.

**Zone-V:** (3500-5500 ft) Subtropical Hill Savannah is an area of tall grasses, black wood button tree, emblic myrobalam, wild olive 'sal' tree, wild date, teak etc. The Paliyans come to these areas during summer to herd the cattle of plain people.

They also come here for the annual burning of these grasses. In this region Paliyans find the palms, utilized for Sago. Puri thinks formerly there were dense forests in this zones.

**Zone-VI:** (Above 5000 ft) These are grasslands interspersed with temperate forest. Though other herders of the plain reach this area, the temperature here is near freezing and hence Paliyans rarely visit this area as they lack proper protection.

## PALIYANS RELATIONSHIP WITH ENVIRONMENT

The Paliyans were almost a nomadic tribe earlier although their movement was always confined to a limited territory within which they moved from time to time in small groups or an individual shifting to neighbouring village. But at present their movements are still further restricted. While some are fully settled, others are semi nomadic. In either case, they have very close relationship with their immediate surroundings. With a group of people in Srivilliputtur area, I tried to know their relationship with the surroundings. They informed the names of particular places, the significance of each and their function.

As the Paliyans here have a stronger dependence on the forest, which usually starts from the foothills of the mountains, an attempt has been made to study how they identified and distinguished it. In the sample village of Shenbagathope of Srivilliputtur, an effort was made to find out how they name different regions of the forest and mountains. They call '*Sircum, parai'* to a portion of the mountain, which has a good algae growth on it. *'Parai'* is a protion of the mountain. *'Tani parai'* is the name of a place where there is a constant source of water. Again some places are associated with deities. *'Midi Adicha parai'* is the name of the *Parai* where the traditional goddess 'Pooma Devi' exist. Every year they celebrate the function here in the '*Kartikai' Masam* i.e during the month of November–December. On this occasion, all the Paliyans in the hamlet start eating tubers only after these are offered to 'Pooma Devi'. This is akin to the religious practices of the neighbouring Hindus who celebrate *'Pongal'* as thanks giving ceremony for the success in their harvest. The idea of worshipping a deity before exploiting any resources might have been borrowed from the Hindus through the interference of contractors and other caste people. The Paliyans validate their arguments by saying that just before the honey collection season the contractors encourage the Paliyans to pray God before commencement of the mission. Some places are named after certain important events. '*Rajam Parai'* is named as where Raja or Raju use to come to this mountain. '*Kalli Parai'* is one in which there are lots of *Kalli* plants.

Similarly, hillocks are given different names. Some of these hillocks have a folk story behind it. ' *Tavala Mottai*' is the name of a hillock where there is a stone at the hilltop, which looks like an animal. They say a live crocodile got formed into stone at this place. '*Vellan Kal Mottai'* is the hillock where the first sunrise is seen. '*Pai Malai Mottai'* is another hillock where there is an '*Amman kovil*' (Temple of Mother Goddess).

Legend goes that the mother goddess of this temple is so powerful that she foiled the effort of a missionary priest who was credited with lifting any huge rock with ease, but failed to lift the stone near the shrine. '*Kalli Parai'* is another portion of the mountain where there are many trees by the name '*Kalli*'. Occasionally traders come here to collect the milk of this plant. Similarly they call the forest by different names. '*Adaippan Kadu'* is one in which there is thick forest with lot of animals. *'Karasu Kadu'* is one where there are thorny trees and small trees. Again, *'Moongaluthu Kadu'* is one where (in the forest) there is a god by the same name. They call '*Metan Kadu'* to a portion in which there are '*Pallam*' (stone) in all the sides. '*Unapatti Kadu'* is a portion of the mountain which is filled with creeper by the same name.

'*Kavu'* means forest. They call a portion of the forest as '*Sembavalli Kavu'* where there are excess of *Sembavalli* flowers. '*Kuluru Atti Kavu'* is the portion where one starts shivering. '*Chitaatuo Kavu'* is one where huge trees are present forming a close network in the forest. This place is filled with small streamlets. And interestingly there is a stone in this place, which is called '*Soolai Kallu*' where a bird of the same name comes to the stone early in the morning by 6 am. And again in the evening to create flutter. Their sound is so melodious that it helps them to wake up early in the morning and again relax in the evening. '*Pandala Kavu*' is a portion in the forest where the trees form a canopy like a Pandal.

'*Voda Medu*' is that area where the forest is very grassy and it is not possible even to enter. '*Upppu Medu*' is a spot where the soil is salty in taste.

They try to name the streams based on its location or sometimes on some distinguishing features around it. '*Paliyan Parai Odai*' is that portion of the stream near to which there is a hillock and there are some stone caves. Paliyans state that there ancestors lived in that spot. So as the stream flows on the foot of the hillock the stream in that portion is so called. '*Sevaru Kani Odai*' is a canal which has the stream in the style of steps in the rocky portion. '*Orai Keni Odai*' is that portion of stream where it is risky to walk over or to take bath, as there are circular pits inside it. There live the water snakes. '*Manga Ottu Odai*' gets its name due to excess mango plants in this region. The name '*Molai Keni Odai*' has come about due to the fact that water flows from a cave like structure. There are other '*Odai's*' such as '*Vannan odai*' '*Siddattu Odai*' and '*Peikal odai*' around their habitat.

There are other places in the forest, which are known by their peculiar characteristics. For example '*Siratan Keni*' is a spot where people in the earlier days have drunk water with the help of coconut shell. '*Maradadi*' is another water spot (*oothu*) behind *Alagar Kovil* (temple dedicated to *Alagar*) where the water is perennial. '*Vella Oothu*' is another water point where the water is perennial and it looks white. '*Kadaka Sanai*' is a place where the perennial Oothu is located in the midst of galinut grove. '*Pai keenar*' is a portion of the stream that has obtained its name from the deep black colour of the water. '*Pallu Chunnai*' is a spot where a *kodi* or creeper by this name is predominantly available. '*Vandi parai*' name has come about due to the fact that people who want to go to the hill top leave behind their vehicle at this place.

Similarly '*Kuruvi Kanel*' is a spot of the forest which is very thick. The speciality of the place is that the birds are abundantly seen in this place. '*Kuttu Kallu*' is a place where the hillock is stony and pointed. The people go to this place for collection of honey. Interestingly they call the markings installed by the forest department '*Kaiatti*'. '*Veer sedi*' is an area where cattle's are taken for grazing. There is a place where there are white rocks knows as '*Gadda Pallam*'.

The above description shows that the Paliyans spend a good amount of their time outside their home, and enjoy the familiarity and emotional attachment to surrounding areas of their settlement. They have a few shrines inside their settlement and many deities in the surrounding hills and forests, which cannot be identified by outsiders. In their familiar areas they follow the age-old occupations within an area surrounded by caste people who live very near to their habitat. But their very occupation keeps them separate from their surrounding caste people inspite of their close interaction with them.

## BIOTIC RESOURCES OF WESTERN GHATS

In an effort to find out how Paliyans exploited the floral resources they were asked to enumerate the plants around them. They described the plants under three categories; namely '*Chedi*'– (Plant). '*Kodi*' – (Creeper) and '*Maram*' (Tree). They spoke how most of the plants were useful in their daily life and provided source of income for them. These informations were collected on the basis of group interview in both the hamlets. Basically they depend on the plants for the following.

(i) Collection of edible fruits, seeds, greens, tubers etc. for their own consumption's.

(ii) Collection of edible and other minor forest produce for selling purpose.

(iii) Collection of medicinal plants for sales.

(iv) Collection of other minor forest produces for sales in the local market.

Apart from the above, they also spoke about peculiar features of certain plants which cannot be put in any of these categories. For instance a *chedi* by the name '*Nal Tulasi*' is used as mosquito repellent. They believe that a plant by the name '*Mani Kulliki/Salunga Chedi*' when touched by children would invite fever. Similarly they also believe that a plant by the name '*Malai Vasam/Malai Mungil*' when cut, burnt and dumped in the stream, welcomes rain. They also state that '*Kattamanakku*' can be used for fencing as it has good growth.

Among trees the fruits of '*Villai*' is used for removing evil eye. Children for playing use '*Katate Maram*' fruit.

In the creepers *(kodi)* category, the juice of the plant '*Tee Kodi'* is believed to create a burning sensation in the body. And the flower of '*Siru Pani Poo Kodi'* when exchanged between husband and wife with laughter then both would remain happy always. On the contrary if the flower is exchanged between them without any emotion then it is likely to cause anger.

It is believed that the creeper '*Pali Kodi'* can swell the body of an individual if it is crossed. This is considered as revenge by the plant for not showing respect towards it. Similarly there is another *Kodi* (creeper) by the name '*Kutu Puzuchi Kodi'* whose fruit they believe, when taken by ladies leads to menstruation. So ladies do not even touch the plant. The seeds of '*Tamba Tavarui'* is used by children in rubbing it in the floor for generating heat and then it is used as an instrument for touching against any fellow being to give him/ her a shock.

Edible fruits are many, which may be directly eaten or eaten after preparation. Some of these are eaten to quench hunger and in pleasure while they eat some as medicines. The edible aspects of the fruits will be discussed here, ignoring the compartmentalization of these categories.

The fruits of the following trees are eaten.

*'Athi': Ban hinia racemosa*

'Elandai': Ziziphys maurutiana lank

*'Kodi Naval Palam':* Small Sweet fruit of a bush

*'Katu Koyal':* Guava

*'Kodu Kapuli'*:

*'Kadakai': Terminalia Chebula*

*'Mullu Maili Maram':* A small tree fruit

*'Murungai':* Drum Stick, Morings – *Olefera Lank, Morineaka*

*'Navai' (Jamun): Eugenia jambolana*

*'Nellikai-Amla': Emblica Officinalis*

*'Puli':* Tamarind: *Tamarindus Indica*

*'Villa':* Wood apple fruit: *Feroni Vellonia*

In the creeper category, the fruits of the following plants are taken for consumption.

*'Kilachi Kai Kodi'* (the fruit is used for pickles)

*'Kallu Tattan Kodi'*

*'Kuruvi Pavakai'* - *Momordica Charantia*

*'Kovai Kodi'* - *Coccinia indica*

*'Puluchi'* (fruit yielding in *chittarai)*

*'Paluva Pavakai'* - *Momordica Charantia*

*'Siru pani Poo Kodi'*

*'Theva Kodi'*

Moreover the following fruits are also eaten.

*'Aaccam Palam':* A mulberry – like tree (*Guazuma tongentosa*)

*'Allam Palam': Arenga Uightii griff*

*'Itchi Palam'*. May be *Phoenix sylvestris*

*'Kala Palam'* - *Randia Sp.*

*'Karam Palam'* - *Randia Sp.*

*'Karunj joorai Palam'*

*'Kaththalai Palam'* - *Aloe vera*

*'Katu Erikulam Palam'*

*'Katu Manga'* Wild Mango (*Mangifera indica*)

*'Koota Parah Palam'*

*'Maghulam Palam'* - *Manusops elengi*

*'Manjanutti Palam'* - may be *Morinda citrifolia (M. bracteata)*

*'Mittukkam Palam'*

*'Pulam Palam'* - *Kirganelia reticulate*

*'Puli Joorui Palam'*

*'Tumbula Palam'*

*'Udupai Palam'* - *Grewia serrulatai*

*'Unni Palam'* - *Lantana indica*

*'Valukai Palam'* -

*'Virusam Palam'* *Cordia dichotoma*

Paliyans also take the seeds of various plants. These include

*Atalam Kottai / Taalam Kotqai / Vandalam Kottai*

*'Katu Pala Kottai'* - *Artocarpus hetero phyllus*

*'Konguti Kottai'*

*'Mullurai Kottai'*

*'Murungai Kottai' (Moorings – Olefera land, Morineaceae)*

Paliyans take the following greens namely
'*Agati Kirai' (Chinna indam Kirai) - Sesbania grandiflora*
'*Icham Kirai' (c) Phoenix sylvestris* (L) Roxb.
'*Indam Kirai Pterolobium hexapetalum'* (Roth)
'*Injai Kirai* Zingiber Officinate Rose'
'*Kana Mu lagu Kirai' (c)*
'*Katu Murunkai Kirai – wild drum stick'. Moringa Oleifera*
'*Keeva Nelli Kirai' – Phyllanthus fraternus*
'*Kovam kirai'*
'*Kumuti Kirai'*
'*Kuppai Kirai'*
'*Moghili Kirai' - Acacia suma*
'*Mahili Kirai (c )'*
'*Manali Kirai'*
'*Minna Kirai'* - may be *premna serratifolia*
'*Molaga Takali Kirai'*
'*Mosu Mosu Kai Kirai'*
'*Mullai Maram Kirai'*
'*Mulli Kirai' - Amaranthus sp.*
'*Mullu Kirai' (c )* Amarantus spinosus
'*Nai Uruvi Kirai' - Stachy tarpheta indica*
'*Parapu Kirai (c )*
'*Ponnam Kanni Kirai'*
'*Puluka kirai (c )'*
'*Sarana Kirai'*
'*Sarana Kirai'*
'*Siri Kirai' - Suaeda maritime*
'*Thaivallak Kirai'*
'*Toguli Kirai'*
'*Tolli Kirai (c )'*
'*Udumbu Kolupu Kirai (c )'*
'*Vatta Konnai Kirai' (Tree) Macaranga Peltata*

## Medicinal plants

Paliyans not only collect tubers, fruits, green vegetables for their own consumption and other minor forest produce for sale or barter but also collect plants, roots or fruits for medicinal purposes. The income generated through the collection of medicinal plant is note worthy and this has been discussed in detail latter. Basically the Paliyans have the advantage of knowing in which part of the forest each plant grows. Their knowledge about medicinal plant can be categorized into two types. In the first type their knowledge is not limited to the identification of plant and its habitat alone but they also have the ability to prescribe its usage for curing various ailments. In the second type they can identify and collect the plant as and when requested by their customers. Normally they do not reveal their knowledge about the specific use of these plants. Only when the caste villagers approach the expert person with their problem they try to provide the requisite plants. While the specialist in one village was a Dhevar man married to a Paliyan lady the specialists in the other hamlet, were Paliyans only. A frantic search by outsiders for medicinal plant in the Western Ghats have made the Paliyans more suspicious and confused. While one organization went ahead with collection of medicinal plant for protecting their patent right in favour of tribals another went ahead with their conservation aspect. Still others tried to exploit the knowledge of Paliyan in discovering some new medicinal plants. So the researcher had to gather data slowly and through discussion from time to time. For instance, in the initial stage whenever an attempt was made to find out whether they had any traditional medicine to prevent pregnancy the reply was in the negative. But once, while discussing with a youth about his not having any issue he gave the clue. He said his mother had given the '*Katu Pusinikai*' tuber to his wife to prevent pregnancy forever. And this has resulted in sterility of his wife. In my querry as to why his mother had given such a medicine, which had a stronger reaction his answer was that his mother did not wish them to have any children. While the data was analysed it was observed that he had married to his parallel cousin. This came to light while enquiring about the marriage rules, which are followed by both Paliyans and Non-Paliyans. Paliyans did not reveal these information's immediately. Only when I had developed cordial relation with them they told me about it. On the contrary some non-Paliyans told me "Aadu Madu Onu Paliyan Onu" which means that there was no distinction between cattle's and Paliyans. One non-Paliyan while citing the above case told me that till the other day this boy was calling his wife as "Akka" or elder sister. But after eloping with her he calls her his wife. When I had asked this young man whether he can overcome this permanent sterility of his wife he replied that his mother knows some plant but she is yet to give it. It appears that the mother of this man did not approve the marriage of the couple. Unfortunately during my subsequent visit both the couple and the mother of the man had shifted to a different hamlet.

The above incident made me more curious about the knowledge of the Paliyans regarding the herbal medicines and I tried to make systematic efforts to collect data on ethno medicine. As regards the medicinal plants for common diseases, I found sometimes even small children are aware of some plants and their effects. Though the efficacy of the medicinal plants as stated by them is uncertain yet an attempt has been made to comprehend them. For stomachache the following plants are used.

*Kiradi Kodi Kilangur*: Tuber of *Kiradi* Creeper

*Malai Tangi Veru*: Roots of *Malai Tangi*

'*Peria nangai'*. *Polygala telephoides*

'*Seeri Keeza Kodi'*. (Creeper)

'*Thalai Surali Ver'*. Roots of (may be) *Pandamus tecorius Barkinson*

'*Vayuta Valli Ver':* Roots for stomach ache.

For example, the method of eating *Kiradi Kodi Kilangu* (tubers of Kiradi Kodi) and *Malai Tangi Ver* (Root of *Malai Tangi*) is to make paste of either of them and then mix it with water before taking. In the case of Thalai Surali the root is made into a paste and drunk. Similar method is applied for the root of *Seerkeeza Kodi*. For *Vaitu Kadapu* the milk of the tree *Shenmuruku* is taken. Apart from that the *Pattai* (bark) of *Mullu Kottan* tree is put in water in the night and in the morning it is grinded and taken. For *Vaita alasal* the root of the *Kadupu Chedi* is munched and thrown while drinking the juice of the plant. The plant *Neela Vahai* is used for gas, appetite and excess hiccup. '*Katchan Tarai'* plant is used for hunger while '*Kodi Kalli'* is used as an appetizer for children. And for indigestion of children, *'Pulium perandal'* plant is used. For excess biles, plants such as '*Mavulingi' (may be crataeva nurvela)* and '*Kovara Tangi'* are used. The fruits of '*Kuruvi Pavakai' (Momordica charantia)* are eaten for preventing mucus.

For *irumal* i.e. cold, the fruit of *Katu Goya* is eaten. *Moota nari Chedi* leaves are smeared and its juice is taken for cold. Also the fruit of *Pongai Maram* is tied around the neck for persons suffering form *'Irumal'*. Again similar method is adopted with the plant of *Musu Musukai*. They sometime apply a paste made from the milk of *Erikutam* plant along with *Sunambu* (lime) in the throat region. Apart from the above they also take the leaves of *Tulasi'* plants.

For headache also certain plants are used. The creeper by the name '*Vel Pariti Kodi*' from which leaves are plucked and the milk that is collected during this process is mixed with '*Sunambu*' (lime) and rubbed in the forehead to cure headache. When they suffer from one side headache, they use the '*Virali*' plants. Also they apply a paste made from the turmeric and '*Sunambu*' (lime) in their forehead.

They also use certain plant as first-aid medicine for persons injured by an '*aruva*' (steel knife) or similar items. The leaves of '*Vettu Pachhalai*' are smashed and smeared with '*Sunambu*' (lime) and applied on the spot of the injury. '*Karupan Kollai*' is used for knife injury. For persons suffering from '*Mullu Kuthu*' (Piercing of thorn) '*Erikilai Chedi Pal*' (may be '*Calotropis*' '*gigantea*' latex) milk is put on that spot. The same treatment can be done with the milk of '*Vellai Erikile*'

For curing fever, they adopt an unique method. In this method '*Sengal*' (brick) is made red hot. Simultaneously '*Katu Nochi*' and '*Nerunji Mullu*' are boiled in hot water in a vessel. The patient is covered with a bed sheet and the vessel containing the boiling herbal water is brought and placed under it. Then the red hot bricks are brought and put into this hot water. This generates steam inside the covered bed sheet. By this process the patient starts sweating profusely and gets cured of fever. In a different kind of fever in which the stomach and body swells for which they take '*Nail Panai Kilangu*'.

The leaves of the plant '*Kattu Kodi*' has the effect of cooling the body. The leaves are simply put in ordinary water and a *halwa* like thing is formed. This is then eaten for body heat. They state that '*Alli Kodi*'. '*Megha Sanjivi*', '*Ana Nerinji, 'Kumunam Kolai*', (probably *cucurbitacea*) '*Pennangani*', '*Nannari* (*Hemidesmus indicus*) '*Vella Adalai*' has also got the same effect.

Plants such as '*Taivalai Kirai*' '*Ponnavarai*' are taken for asthmatic conditions. Also the '*Vella Erikile*' flowers are used for ulcer and asthma problems.

For joint pain '*Palai Maram Pattai*' is boiled in water after powdering it. After sometime a pasty substance is formed. This is then spread in the affected area and thus the pain is relieved. The milk obtained from the '*Kalli*' (may be species of 'ficus') plant and '*Thirugu Kalli*' plants are used for relief of joint pain. Another method of curing the disease is to grind the root of '*Naga Tali*' and apply it in the joints. '*Katatti*' leaves are also used for same purpose. The plant '*Katchan Torai*' is used for joint and backbone pain.

The leaves of '*Mookoti Kodi'* are smashed and applied for summer boils. '*Katu' 'Karna 'Kilanga'* is helpful in curing '*Moolatu*' i.e. piles related problems.

The '*Pattai'* of '*Chittai Maram'* is useful for curing '*sorangu'*. The method involves grinding the *'Pattai'* and putting it on '*Sorangu* '. Also the leaves of '*veppilla'* tree is used for curing this disease. The plant by the name '*Kuppa meri* is used for curing scabbies. The plant '*Palam Pasi'* is useful in curing '*leukoderma'*.

For teeth problems such as teeth decay the tuber of '*Pallu Valli' 'Kalangu'* is used. The tuber is burnt and munched at the appropriate spot. This is believed to prevent teeth decay. For ear pain '*Seghitu Olai'* plant is useful.

There are also plants which they claim can be used as alternative modern medicines. For dog bite, '*Nai Uruvi'* leaves and ' *Sunambu*' are grinded and applied on the dog bitten spots. Both '*Peria Nangai*' and '*Mulaga Nangai'* are used for snake bites. Also *'Ariya Nangai'* and '*Pavala Nangai'* are useful for the same problem. The method of eating it involves powdering the whole plant. The powdered plant is then mixed with hot milk or water. They state that this may be taken during any poisonous bite. Also another plant that can be used here is '*Mokara Chrnai'*.

'*Vella Araghu'* and *'Kiwanelli'* plant are used for curing '*Manja Kamalai'* (Jaundice).

## Fauna

It is interesting to learn from the local forest Department plan (1994) that Western Ghats, in general, is one of 221 priority areas for the conservation of global biological diversity. This includes 6 endemic mammals, 77 endemic reptiles, 84 endemic amphibians, 4 endemic butterflies and above 1800 endemic plant species. Some of the important mammals found in this part include 'Grizled Giant Squirrel' (*Ratufa maccoura*), Elephant Langur (*Presbyits Johni*), Lion tailed Macaque (*Macaca silinus*.).

Apart from these animals some of the commonly found animals are 'Gaur' (*Bos Gaurus*) 'Sambar' (*Cervus unicolor*) 'Porcupine' (*Hystrix Indica*), 'Nilgiri Langur' (*Presbytis Johnii*) 'Sloth-Bear' (*Melurusus ursinus*) 'Malabar Giant Squirrel' (*Ratufa Indica*) Both common Langur (*Presbytis centellus*) and 'Bonnet Macaque' (*Macaca radiate*) are very common. Similarly wild boar (*sus scrofa*) and Indian hare (*Lipus higricollss*) are in abundance. But barking deer (*Muntiacus muntjak*) mouse deer (*Tragullus meminna)* 'Dhole' (*Cuon Alpinus*), 'Panther' (*Panthera Pardus*) are rare. And 'Chital' (*Axis axis*) Tiger (*Panthera Tigris*) are very rare.

## Technology of the Paliyans and their Subsistence Pattern

As described earlier, Paliyans live in rocky area with thorny bushes without shady trees. During summer the soil becomes hot and there are no trees to give them shelter. The 'Sangam' Poets of Tamil Nadu have described such areas as 'Parai'. Gardener raises the question why the Paliyans prefer to stay in such inhospitable regions when the top of the mountain ranges in Palanis and its contiguous areas are filled with fertile soils that attracts caste people from the plains. He suggests two plausible reasons.

(a) The dry stony inhospitable areas of Palani habitat provides a large number of 'diascorea yams' and greens, small animals, birds, eggs etc. which Paliyans do not exploit fully but allow these consumable resources to waste (Gardner: 1972,1994). This availability of staple food such as 'diascorea yams', 'green leaves', eggs, small birds, small animals etc which provide them with sufficient nutrition for their consumption and attracts them to these apparent infertile areas.

(b) Secondly no outsider ever tries to compete with Paliyans for such tubers which are difficult to dig and tasteless to consume. Free from the competition in such areas Paliyans feel free to follow their style of life without inhibition from outside people who usually look down upon them.

### (i) Technology of food gathering

All ethnographers emphasize that the digging stick is the sole and only implement that they manufacture and utilize in digging 'diascorea yams' which is their staple food. Paliyans prefer to use the wood of '*Envelli*' for making the '*Kombu*' (digging sticks) when that is not available he goes for '*Mulli*' or '*Kurundam*' or '*Sulpurat Kombu*' or '*Maila Kombu*'. These '*Kombus*' are cut and shaped based on the requirement. The length varies from 25 cm to 150cm. Smaller ones are used for scooping the soil laterally. The tip of the '*Kombu*' (sticks) is shown in the fire for better sharpening and preventing splitting. Though digging stick had been their only tool from time immemorial perhaps they cannot give up this tool even today due to some advantages mentioned below:

(a) Digging stick does not cost them any thing.

(b) It is easily available and can be prepared based on the required size in no time hence they can use and throw it.

(c) A metallic tool is likely to damage the tuber and hence the use of wooden stick is preferred.

Besides every Paliyan has his all purposive bill hook which he either purchases or obtains from outsiders by bartering honey or minor forest produce. Sometimes he borrows iron tipped spears from his neighboring caste people. But documents during colonial period refers occasionally that in 'Tirunelveli' Paliyans used bow and arrow. In some places they stealthily used country made guns. These were rather exceptional cases but they very often depend on stones, traps and other such simple devices for hunting animals and birds while fishing is done by poisoning the entire fountain, or pools. The details about hunting, using of traps, snares etc. reflects their simple technology which is dealt in the next chapter.

## (ii) Food gathering of the Paliyans

From time immemorial the staple food of the Paliyans had been various kinds of Dioscorea yams, that they collect throughout the year. Only in serious drought or famine situation, they fall back on the 'Sago trees' (*Caryota Urens*) and make gruels to satisfy their hunger. But by and large food habits of the Paliyans are changing rapidly. Yet in interior areas yams and tubers are still the staple food. In most other villages where Paliyans are directly influenced by market economy, the food is occasionally supplemented by diascorea yams. This is mostly done when they have no money. Thus the old habit still continue and digging of roots and tubers are associated with collection of minor forest produce.

Morning scene is interesting in a Paliyan village when all the villagers are awake and in their relaxed mood, the women are busy in their household chores such as cooking food, bringing water from the nearby stream of pool and cleaning the surrounding. Adult males and old men just relax or continue their gossip. By nine 'O Clock they take some food, either cooked in the morning or some leftover food from the previous night. Some people go straight to the jungle. Unlike earlier habits no longer they go in small groups at the same time. The domestic labour group, mostly the husband and wife, go together to the forest for their foraging activities. Mother carries small babies while those who can walk are left in charge of grand parents or some elderly siblings. The village becomes deserted soon, with a few old men or women and some people who prefer to relax after collection of good quantity of minor forest produce in the previous day or days.

The small labour groups which are formed on the basis of mutual friendship or close relationship may go together upto certain point in the forest. Then, one or two persons or couples may go in search of the yams or other desired item. Yams are usually found in between granite boulders or some portion of a tuber may be under a rock. In such situations they have to work strenuously when the spouse or some friend may help in the operation. Women usually bring their digging stick from their homes. They straighten the bends of the stick in the previous night by showing it to the fire. But males prepare their digging stick at site when they require this and throw it away while returning home. In most of the times the digging is difficult due to hard soil and from time to time they sharpen the blunted points with the help of their bill hook. The work continues till the entire tuber is brought out from three or four feet below the top soil. After digging out, they cut the upper portion of the yam and again bury it in the pit with loose soil around the tuber for its regeneration irrespective of who ever harvested it. This grows later. Thus foragers indirectly conserve their principle dioscorea yams which provide them with their staple food.

The digging operation is always accompanied with merry making and jokes at intervals when they are engaged in searching various items such as fruits or plums or other delicacies which they consume directly. Along with such collection they also collect greens etc. and carry out the collection of minor forest produce when there is demand by the agents of the forest contractors.

Sometimes, in the mid-day they make fire out of leaves and dried branches of a tree to prepare food. In between, if possible, they catch small rabbits etc for consumption. After their lunch they rest for a while, some may even sleep and again start their foraging activity. In the afternoon, they collect the firewood, which they carry home along with other items such as tender bamboo shoots, eggs, seeds and minor forest produce whatever come on the way.

### (iii) Different kinds of tubers consumed by the Paliyans

Mostly, three kinds of diascorea yams used by them (a) *'Vetale valli Kizangu' (Dioscorea opppostifolia), (b) "Mullu Valli Kizhangu' (Dioscorea pentaphylla), (c ) 'Nuulam patai' or 'caipam' (dioscroea tomentosa).*

(a) **Dioscorea oppositifolia:** This tuber is available round the year in the forest range. Though the leaves dry off in the summer months yet it can be recognized from the nearby dried creeper. The plant is seen in the rocky terrain at a height varying from 300-1250 meters. It can grow very deep and has the advantage of giving large tuber.

(b) **Dioscorea pentaphylla:** This tuber is preferred by most of the people due to the fact that this tuber is very large. It is stated that the local people also show greater desire for the consumption of this tuber and thus purchase the same. A single plant can yield ½ kg to 1kg of tuber. It mainly grows on the lowest slope of the mountain.

(c) **'Dioscorea tomentosa:** This tuber can be obtained at a lesser depth and grows at a height of 300-1250 mts. The tuber grows in the summer and hence available for consumption during rainy season. The tuber is said to have some medicinal value. While the field work was in progress during the summer months, caste people of the neighbouring villages came to Paliyan hamlet to request them to fetch them this tuber. The tuber is said to be a curative for the disease of 'piles'.

Though the above mentioned yams forms the major varieties of yam consumption, the Paliyans also consume few varieties of other yams. These are eaten in various ways. While doing the field work, I observed one night a lady, after preparing the dinner put few pieces of tuber in the burning, charcoal. After sometimes, she removed them and gave me a piece to eat. They also eat it by boiling them in water. Another interesting method of eating the tuber is by stone baking. There are variations in stone baking of the tuber. And it is reported that this method is adopted while they are engaged in the annual root offering festival for the protecting god. This festival is organized collectively by the Paliyans of a specific hamlet.

The Paliyans also state about the consumption of the stem of certain plants belonging to the 'cycas' family.

The above description indicates how the Paliyans living in rain-shadow areas of Western Ghats, eke out their living in a dry, hot climate using their simple technology.

The Paliyans, scattered in an extensive area, gather sufficient food stuff for their own consumption with their simple digging stick and all purposive bill hook. The digging sticks are available in the Paliyan habitat easily. Both males and females exploit the resources of nature in this rather infertile area which contains large amount of natural resources. These can not be fully utilized by the Paliyans and after consuming a small portion of natural resources the Paliyans waste a large amount which remains untapped or allowed to waste. Paliyans display their ingenuity in collecting honey from high altitudes with the help of their vine ladders. In this risky operation they take the help of most reliable affinal kins i.e either the brother-in-law (wife's brother) or father-in-law. For collection of honey they have sufficient knowledge of the behaviour of various types of bees and also know how to avoid their stings while floating in the mid air.

In hunting small games they exhibit their special skills in catching or killing rabbits, rats, squirrels, wild fowl, various types of birds with the help of simple traps or snares. The only big animals that they hunt are the wild pigs and deer either by driving them to dangerous rocks or ambushes where they can be attacked and killed easily. The mechanism of killing big games with the help of marvelous devices made with big rocks, smashes the animals when it touches the pole that support it. Further, most often, while hunting the animals they follow the trails of animals which are wounded either by a poacher or by a predatory animal. In fishing also, their way of poisoning the streams or pools is also very simple but it helps in collecting fish almost without making any strenuous efforts.

Thus, Paliyans by choosing to live in inhospitable, infertile hot, dry, stony land situated in rain shadow areas in low altitude manifest their age-old experience and intelligence. Living in such habitat, they protect themselves from biting cold of higher altitude and in the water logging marshy valleys of high altitude, which breeds mosquitoes and invites malaria. Finally they lead a contended life in their surrounding natural landscape being free from competition from plains people who cannot lead such hazardous and painstaking life and can not push them away from their present ecological set up. They follow their simple technology. Without disturbing the ecological balance, they adapt themselves to nature and add to the natural growth of flora and fauna, which would be later discussed.

## SECTION – II

## SOCIAL STRUCTURE OF THE PALIYANS

In the present study the author has selected two Paliyan villages for the intensive analysis which seem to differ to some extent in their ecology, economy and in respect of developmental process. Each village is differently influenced though both are substantially exposed to outside influences. Yet, both have some common elements in their core culture.

The village Palamalai which is situated at a height of 4000 ft. in the upper Palani hills, is in a plain valley with fertile soil fully controlled by Jesuite Christian Missionaries since the beginning of twentieth century. It is located within the missionary estate which extends over a vast area measuring more than 250 acres (approximately). The village is now within the Anna Dindigal district. It is on the side of the *pucca* motorable road which leads to Kodai Kanal known to be one of the most important tourist centre and summer resort in Tamil Nadu.

The second village Shenbagathope is situated at a distance of ten km from Srivilliputhur and belong to Kamarajar district of Tamilnadu. The village is situated on a plain land not far from the foot of the hills of Western Ghat. It is surrounded by agricultural fields, fruit orchards, etc. while the fruit bearing trees of the temple of 'Andala' of Srivilliputhur flanks the other half of the village. There are a number of Hindu temples within the grove of Andala temple. 'Alaghar Koil', a pilgrim centre, is situated in the middle of hills. Excepting the tea shops and few families; the village Shenbagathope is fully occupied by Paliyans who still retain a large number of traditional features of Paliyan culture. They are also greatly influenced by surrounding Hindu culture.

Environmental factors differ widely between the two villages. Village of Palamalai is under the influence of missionary fathers and they are relatively more advanced in education, agriculture, and in modernization process. The Paliyans of Shenbagathope retain much of their traditional foraging practice and quintessence of their traditional culture. With the help of data collected from these two villages we have tried to depict the Paliyan social structure and analysed how ecology influences the Paliyan social structure.

## SOCIAL STRUCTURE OF THE PALIYANS

In any society, family is the basic and indispensable unit on which it is built up or organized. But before discussing the family as a basic unit of the social structure, marriage as an institution which is instrumental for the formation of the family will be discussed in this section to understand the nature of family in the Paliyan society.

**Marriage:** Though the Paliyans are influenced by the customs and traditions of the caste people of the plains area at present, the institutions of marriage and family of the Paliyans differ significantly from those of the caste people of the plains. This occurs despite the fact that the former while trying to adopt the institutions of the marriage and family of the plains people, retain the attributes of their own institutions due to their peculiar ecological factor and their own archaic traditions and institutions.

According to Gardner, "what they call '*kalyanam*', a marriage, is simply a union that is public". When a girl attains puberty, she has two options. Either she runs away with her fiancée or she gets married through arranged marriage. The former is more popular. Here, once a boy and a girl like each other, they wait for an opportunity to run away from the hamlet to the forest areas. After spending a night or more outside the hamlet they return home. Both are accepted as husband and wife by their fellow men.

**Rituals of arranged marriage:** Though the ceremonies of marriage are very simple, we see some regional variations in the marriage procedure at different places. In lower Palani hills the marriage is fixed by exchange of betel leaves and areca nuts. The date is fixed for the marriage. On this day the bride-groom along with his kin members proceeds to the bride's house with new saree, blouse, cosmetics etc. On arriving at the bride's place, the girl is called in public. Then she is asked to wear the saree etc. and come out if she is interested in the alliance. If she is not, then she does not accept these dresses in public and the proposal breaks off. Once she agrees to the proposal, the groom's party come home with the bride. The marriage takes place on small pandal created for the purpose by tying the '*thali*' on the bride's neck. A common feast is given to the near and dear ones with which the function culminates.

In another part of the Palani hills a Mannadiar' (a cultivating caste who are neighbours of Paliyans for four to five centuries) is requested to officiate the marriage. The Paliyans who have embraced Christianity have to undergo the marriage rites twice if their spouse is from the Hindu community i.e. marriage ritual is performed both in Hindu and Christain style. The traditional way of marriage is organized in the girl's house and the feast is organized at her place. Afterr returning back to his village on a convenient day, the ceremony is undertaken in the local church.

There is yet another pattern of arranged marriage. They bring a girl home and bring her up as the family member. If she accepts the proposed youth the marriage is celebrated. Otherwise, she is free to choose some one else and can run away with him.

**Divorce:** The longivities of the marriage ties depend solely on both the partners. Separation is very simple and therefore it is easy to see divorce as a common phenomena. Marriage breaks off when either of the partners have developed intimacy with a person other than the spouse.

When a person is not interested with the present spouse, he or she makes love with some one else and establishes the marital ties. Both run to a forest and stay there for a day or two. Then they return to their village so that all know that the previous relationship is ruptured and that they have established a new partnership in their marital affair. In such situations neither the man nor the woman is punished for violation of their old marital ties and establishment of a new relationship.

### Case Study – I

In a hamlet at lower Palini hills a Paliyan named 'Andal' had to abandon his wife when differences arose with her. As she wanted to marry a person with whom she had developed intimacy, she got the consent only after Andal denounced his wife and declared it openly that some one can get married to his wife. The Paliyans in the village felt that this was necessary to prevent any further complicacy. The demand by Andal over his wife or children will be invalid if it is claimed in future.

As marriage occurs more than once among the Paliyans, one can easily find different complex situations taking place. Age is not a factor between the spouses to be taken into account. In a hamlet a man at age of sixty or above got married for the third time to a girl of around sixteen years of age. In the same village a youth aged around 25 years walked away with the wife of a fellow Paliyan. She was older than this youth by 4-5 years.

Though at any given time a man is paired up with one woman, there are stray instances where a man has more than one wife. Marriage outside Paliyan tribe is probably on the increase in the recent days. Many factors may be responsible for such changing trend.

### Case Study No–II

In a hamlet, a Paliyan boy was groomed up by a voluntary organization to be a future leader of the tribe. He is literate and is accustomed with handling of day to day affairs of his village and in the local surrounding area. He fell in love with a caste girl despite his marriage to a Paliyan. Fearing resistance from the village community they eloped from the village. After 2/3 years they returned to the village and were accepted by the community.

### Case Study–III

In another hamlet, a Naicker girl ran away with a Paliyan youth immediately after getting married to a fellow Naicker boy. Though her present husband is her lovers elder brother, yet she had to accept him under difficult situation. After running away from her Naicker husband's house she first met her lovers brother and she had to runway with him as she was apprehensive of being caught.

One distinguishing feature to be noted is that in most of the hamlets taken up for study, one invariably can find a Dhevar man married to a Paliyan lady. It is important to note that in most of the places where the Paliyans are illiterate and have relatively closer attachment with the forest environment, these 'Dhevar' men married to Paliyan ladies and act as intermediator between Paliyans and outside world. Paliyans accept any caste Hindus whom they consider higher to their social status. One reason of such marriages between 'Dhevars' and Paliyans could be the significant presence of 'Dhevars' in the plains area and proximity to their habitat. They do not accept relationship with any one belonging to the community whom they consider lower than their status and those who are considered untouchable in their local area. There are instances to show the increasing trend in violation of these caste norms also.

## Pedogamy

The next peculiar and scornful institution in Paliyan marriage is the existence of pedogamy or the parent child marriage. It occurs in two different situations. In first situation a non-related person is brought up as a child. But the parent child relationship, in course of time transforms into husband wife relationship.

The other form of such marriage is to bring up a stepson or a step-daughter as ones own child and marry him or her in course of time when the child grows to a full-fledged man or woman. According to Gardner occurrence of such marriage is as high as 10 per cent (16 out of 155 cases of marriage) in his field work villages. But, it seems, such an institution is becoming obsolete gradually now a days, with greater contact with outsiders who ridicule such institution.

Commenting on this type of marriage, Gardner suggests, "Informants expressed no interest in the latent economic value of these arrangements, that is as insurance for the aged, but then planned dependence is antithetical to Paliyan mores" (Gardner: 1965: 17).

Gardner gives two reasons for such peculiar marriage arrangement, viz; sexual satisfaction and greater durability of marriage when it is built upon the foundation of a parent child relationship.

Though the above argument of Gardner may appear plausible we can further explore the tribal social situation in an inhospitable ecology where they lead a rather semi-nomadic social life. In a tribal society, where there is no joint family system nor any organization to take care of the aged or the destitute, they are confined to their isolated inhospitable ecosystem, with strenuous life. Each individual has to rely more or less on the spouse for carrying such extra burden. In such a situation a marriage through pedogamy may lead to greater care and ensures greater social support.

Gardner has summarized the distinguishing features of Paliyan marriage as detailed below:

First, the Paliyan considers a union, even of a short duration, to be marriage if it is of a public nature. Second, relation between the step parent and step child or adopted parent and adopted child even where parent child kin terms are used, frequently evolve slowly into marriage relationships. There is no ceremonialised point at which the role changes are presumed to occur. Third, most marriages involve residential, sexual, procreative and economic sharing and co-operation. Although a number of unions, which are, termed marriage lack sharing in one or more of these spheres, this has the nature of being a temporary or substitute arrangement. Fourth, a good deal of individualism is retained by spouse particularly in the sphere of authority. Marriage is based upon egalitarianism, breach of which by either party is ground for separation [*Gardner: 1965:20-21*].

At present the usage of ‘pedogamy’ is declining very fast. In both the villages of Palamalai and Shenbagathope I could not come across any such case of ‘pedogamy’. But in the village Shenbagathope. I heard a rumour regarding a young boy having sex relation with an old lady. But no such case is considered as *‘Kalyanam’* (marriage). With closer association with caste Hindu society, the spread of Christianity and interaction with governmental and non-governmental welfare agencies the institution of ‘pedogamy’ is gradually declining.

On the contrary, higher divorce rate still persists in Paliyan society though the neighbouring castes usually give an exaggerated picture on such issue. The table No. 2.1 will indicate the frequency of marriage of each individual in the two base villages.

Another peculiar feature of the Paliyan divorce is that a spouse may divorce his/her partner and may have one or more relationships with some new persons. But in course of time if the former partners wish to come back, they can be again reunited with the former divorced spouse and can lead their marital relationship afresh.

**Table 2.1: Frequency of marriage in Paliyans.**

| *Name of the village* | *Never Married* | | | *Once married* | | | *Twice married* | | | *Thrice married* | | | *Four times married* | | | *Grand Total* |
|---|---|---|---|---|---|---|---|---|---|---|---|---|---|---|---|---|
| | M | F | Total | M | F | Total | M | F | Total | M | F | Total | M | F | Total | |
| Shenbagathope | 5 | 8 | 13 | 8 | 11 | 19 | 6 | 5 | 11 | 2 | 2 | 4 | 0 | 1 | 1 | 48 |
| | | | 27% | | | 40% | | | 23% | | | 8% | | 2% | | 100% |
| Palamalai | 31 | 16 | 47 | 27 | 22 | 49 | 3 | 2 | 5 | 0 | 0 | 0 | 0 | 0 | 0 | 101 |

**Analysis of the table:** In the village Shenbagathope 27% of the Paliyans are unmarried while in the village Palamalai it is 47%. Individuals married once in Shenbagathope is 40% while in Palamalai it is 49% only. Twice married persons in Shenbagathope is 23% while in Palamalai it is merely 5%. Thrice married individuals in Shenbagathope is 8% and four times married in Shenbagathope is only one case or 2%. In Shenbagathope 10% of people are married either three times or four times while in Palamalai such category of people are non-existent. This indicates that people of Shenbagathope have higher frequency on more than one marriages. In case of Palamalai twice married people constitute 4.5% only.

## Units of social structure

**Family:** As soon as a Paliyan youth marries he sets up his new family which is usually a nuclear one. Types of families found in Shenbagathope and Palamalai are presented below based on Kolenda's classification. Table no. 2.2 shows different types of families found in two sampled villages.

### CHART NO. 1 TYPES OF FAMILIES IN SHENBAGATHOPE AND PALAMALAI AS PER KOLENDA'S CLASSIFICATION

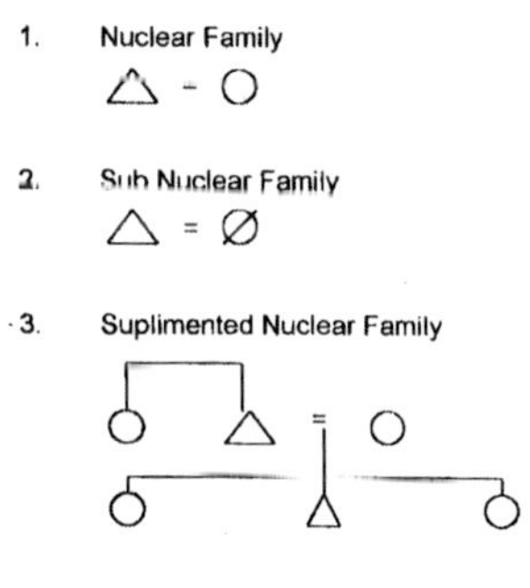

4. Suplimented Lineal Joint Family

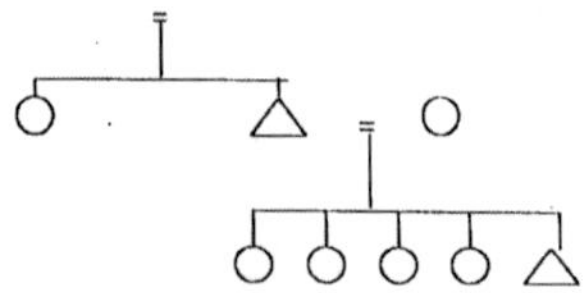

5. Linear Collateral Family

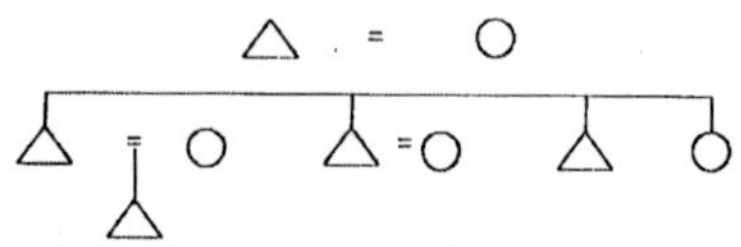

**Table: 2.2 Types of families in Shenbagathope & Palamalai**

| *Name of the village* | *Nuclear* | *Sub-Nuclear* | *Supplemented nuclear* | *Lineal collateral joint family* | *Supplemented lineal joint families* | *Misc* | *Total* |
|---|---|---|---|---|---|---|---|
| Shenbaga-thope | 15 (83%) | 2 (11%) | - | - | - | 1 (6%) | 18 (100%) |
| Palamalai | 14 (70%) | 1 (5%) | 3 (15%) | 1 (5%) | 1 (5%) | - | 20 (100%) |

As per Kolenda, nuclear family comprises of married couples with or without their unmarried children. The above table shows that the maximum number of families belong to the nuclear family i.e 83% in Shenbagathope and 70% in Palamalai.

Sub-Nulcear Family: A sub-nuclear family is a fragment of former nuclear family. In Shenbagathope two (11.1%) such families and in Palamalai one (5%) families are found.

Supplemented Nuclear family :- It consists of a nuclear family plus one or more unmarried, separated or widowed relatives of the parents, other than their unmarried children. In Shenbagathope there is no supplemented nuclear family while in Palamalai there are three (16%) such families. Lineal Collateral Joint Family consists of three or more couples linked lineally and collaterally. Only in Palamalai one (5.3%) such family is existing.

Supplemented Lineal Joint Family consists of a lineal joint family plus unmarried, divorcees or widowed relatives who do not belong to either of the lineally linked nuclear families. In the case of Palamalai hamlet only one (5%) such family is found.

## Interpersonal relationship within family

(a) **Husband-wife relationship:** This is most vital relationship. In Paliyan society each can lead economic independence, and therefore egalitarianism is most often conspicuous. Whenever the husband tries to exercise authority over his spouse it may lead to severe conflict which may end in separation or divorce. Most often the husband is jealous and suspicious about his wife lest she makes love with any other person. There is no strict division of labour amongst the husband and wife, Although a woman is engaged in collecting leaves, fruits or digging tubers for consumption, she is not allowed to join hunting, working in a plantation or collecting honey. Since it is an egalitarian society husband and wife cooperate in most of the house hold works and social activities. But feeding and taking care of the children is the most important function of the mother. The siblings or grand parents may look after the young children. After a certain age, the children can live in any close relation's house for days together. The father is usually affectionate towards his off spring but forgets about the children once separation is held. If the children are very young it is the responsibility of the mother to care young ones. But after a certain age when one is economically independent the child can decide with whom he or she can stay. On the whole a woman is in a much better situation in a Paliyan family and enjoys all the independence to continue or serve marital ties.

(b) **Brother-Brother relationship:** Though brothers before marriage cooperate in all family responsibilities, after marriage, the married brother sets up his separate family. The married brother is some times jealous of his brothers who he thinks may seduce his wife or both can elope. Thus brother-brother relationship is not that sweet always.

(c) **Brother-Sister relationship:** This relationship is unusually strong and a brother whether younger or elder always tries to help his sister in all situations. Whenever a woman is separated or divorced, brother tries to take responsibility of the sister and helps her with earnest sincerity.

In socialization of young ones, all in the family cooperate and take responsibility of caring them when the mother is not free. The children are brought up with love and affection.

## Family and house hold of the Paliyans

The Paliyans still lead a band life to certain extent when they go to interior forest or high altitudes far off from their village. At present they are undergoing a process of transition. In the social life, the family or household plays the most crucial role. As the Paliyan marriage is very brittle, the family of the Paliyans is seriously affected by such marital disruption and undergoes changes.

When a young man marries, he sets up his family with his spouse. If the mother is a widow or the father is a widower and if the siblings are not economically independent they may continue to live with the new couple. By and large the family is nuclear or sub-nuclear or at best supplemented nuclear. In case there is no couple which could form the nucleus, the unmarried siblings or widowed parent with unmarried children may form a house hold. In a family, satisfaction of biological needs, bearing and rearing of children, economic cooperation and education of the young ones are important functions. In a nuclear family of the Paliyans 1st and 2nd functions are performed indispensably while economic cooperation is some times not required. Both the spouses are usually economically self sufficient as each can individually exploit the forest resources for one's subsistence.

**Kinship net work:** The family is the basic unit while extended kinship and kindred are the next important unit of social interaction. After marriage, the couple may reside near to the kinsmen of the grooms. But if there is greater advantage in living near the bride's extended kingroup. (such as MFP rich forest area, or if the bride's father is a leader of the village, etc.) the couple decide to stay with her extended kingroup. Later they are free to change their residence as per convenience. In Paliyan society there is constant imperceptible movement of families or kin members from one camp to another according to their convenience. In life cycle rituals particularly in birth, marriage or death the kindreds unite and carry the functions jointly. In each such occasion persons occupying particular status play the role associated with it that facilitates not only smooth management of particular function but also brings closer integration to the whole community.

**Village and band life:** The Paliyans have more than one place of residence. Previously the tribe used to stay almost ten months in the open camps and somehow managed two months in caves or in readymade houses or temporary huts, made out of locally available grass, bamboo and bushes or with branches of trees. But at present, they lead either sedentary life or semi-nomadic life with a permanent settlement in a particular village. The Paliyans now live in their own villages occasionally accompanied by a few outsiders or plains people. On the contrary, a small group of Paliyan families (say 5 or 6 families) may form a hamlet and live very close to a particular village where caste people constitute the majority. In these villages or hamlets Paliyans form a wellknit integrated community although they maintain their ethnic identity.

The present Paliyan family or household has some unique features. Since each family is related to every other family of the village or settlement either through kinship or affinal relationship no one feels uneasy living with any other family in the village. But every adult person living with any other family, provides the required amount of rice towards his own consumption to his host or the owner of the house (who is also one of his relations). Therefore a Paliyan never feels isolated in a village if he has no house. These people are either unmarried youths or divorcees who have no regular family. They apply this practice for certain period until they set up their own family.

From the childhood the Paliyans are associated with families other than their own which belongs to either a consanguinal kin or an affinal relation 'Sondakar'. This early socialization process makes a person quite free when he lives with others in the village. Paliyan houses are either his own or provided by the government. Most of these are small single roomed houses with a small partition kept apart for cooking. With such close proximity the guest becomes competitor to gain the attention of the host's spouse. With the slightest indication of such intimacy the host becomes suspicious and his marital relationship with his spouse becomes ruptured. Under these circumstances the bitterness continues until the wife elopes with the guest or the host divorces his spouse for her intimacy with the temporary guest or any other near relation. Such disturbed family life at times lead to social conflict and as a solution the emerging couple usually leave the house and stay in another settlement or forest camp. Although in a village or settlement each member feels the psychological identity with the village community, or a family, yet if the situation arises, an individual can migrate to any nearby village or settlement, which may be extension of the first one. Thus in a Paliyan village or settlement, there is always a close kinship and affinal network which generates a sense of identity within the village. The village solidarity or psychological feeling of belonging to the community is sometimes ruptured due to intra familial conflicts. Since the disturbed family changes its residential location, the village peace is not seriously affected nor grows unwildy. Small villages or settlements maintain greater solidarity.

Gardner describes the Paliyans as a non-aggressive people. In true sense since they were partly nomadic and spend most of their time outside the village within the forest environment, they were non-aggressive with outsiders. Within the village in settled life, they have situations for conflict but through their slow migration or movement they avoid the unpleasant situation. Since a Paliyan woman has the full freedom to divorce her partner and can choose a fresh partner there is no perpetual or long standing conflict, which could affect the village solidarity seriously. Due to such freedom of women and tolerance of the Paliyans they have wonderfully adjusted within their society. These habits have developed in Paliyan society because the spouse of each person is a kin member and developed cordiality and close interaction from birth.

The Paliyans have very close relationship with outside castes who are their neighbours. This is quite evident from the following situations.

1. The Paliyans never hesitate to give their women in marriage to a higher caste non-Paliyan. In Shenbagatope a 'Mudaliar' man had married to a Paliyan woman and had a son to him. In Pilavakal dam settlement a Dhevar married a Paliyan woman and became a regular member of that village.
2. Some Paliyans have developed friendship with caste people. They send fruits and other gifts collected from the forest for them regularly.
3. There are folk tales which clearly show the deep rooted relation of the Paliyans with their neighbouring caste communities (Mohanty, 1995: 154-159).

In the following discussions we have shown how the Paliyan culture and social structure are greatly influenced by the ecological factors in both the villages. In these villages not only environmental factors are responsible for their social change but other elements such as influence of historical factors, adoption of innovations through exogenous agencies and the influence of surrounding people are also equally responsible for the formation of their culture and social structure.

The documents of the 19th century indicate that the Paliyans were in the state of nature and their culture was adapted to their environmental situation. Although the culture and social structure of the Paliyans retain their basic features, there is marked difference in the process of social change in the two villages.

Towards the end of nineteenth century there were significant historical developments that accelerated the process of social change. Although four or five centuries back there was slow migration of plains people to the Palani hills, during the colonial period various categories of people moved into it.

In this situation, large-scale deforestation for commercial purposes and the introduction of reservation of forest Act (in 1881) restricted the movement and exploitation of the forest by the Paliyans. The Paliyans, who were physically and socially isolated earlier, came in contact with large number of government officials, petty traders and forest contractors.

During 20th century, along with the inflow of large number of people to the hills, Christian missionaries from U.K., U.S.A and Germany started establishing their churches and this led to spread of Christianity in the region. Simultaneously they started coffee plantations and other fruit orchards in different parts of former Madras Presidency. Other entrepreneurs such as Muslim businessmen, Chettiyars etc. started their plantations which drastically changed the natural environment and affected the food gathering occupation of the Paliyans. The efforts of the Paliyans to adopt to such changing situations has been presented through the discussion on the process of social change in the two sample villages.

In Shenbagathope, they are still adapted to their environmental situation and their subsistence is mainly depending on gleaning economy. The main implements in their food gathering practices are digging stick, billhook (*arruval*) some traps, nets and snares. Excepting bill-hook, which they purchase from the nearby market, other implements are prepared by them from local raw materials. Living within the forest, they had developed the habit of climbing trees and plucking fruits etc. This helped them in collecting minor forest produce.

With the introduction of Government regulations on forest reservation and collection of MFP etc. through contractors, their freedom was curtailed to a great extent and slowly their gleaning practices were reduced. Collection of minor forest produce became more important in their economic life. Inspite of two hundred years of colonial rule and fifty years of post independence era, they are still adapted to their local environment and have not adopted agriculture so far though they are accepting agriculture as their secondary source of income.

The village Palamalai is a big centre of Christian missionary activities. Along with the Chruch, they established a big estate for the cultivation of coffee, fruits such as jack fruit, orange etc. and commercial crops such as black pepper, ginger etc. Majority of the people of the village were Christianized and are controlled or guided by the missionary fathers continuously for over a century. They have almost lost their traditional culture and do not depend on gleaning economy any more. The collection of minor forest produce is minimal. For hunting they borrow the guns and use them conspicuously. Some Paliyans also have agricultural land which they themselves cultivate. They are pursuing diverse economic activities. Paliyans at Palamalai are more exposed to modernization process of the outside world. Both the villages are deviating from traditional way of life due to changing environmental factors and external influence. Yet they are adapting to their local environment in two different ways. Such adaptations of innovations have their impact on their social institutions.

## DISCUSSION

Steward has suggested three approaches for analysis of the method of cultural ecology (1) The analysis of interrelationship of productive technology and environment (2) The behaviour patterns involved in the exploitation of a particular area by means of a particular technology and (3) the procedure to ascertain the extent to which the behaviour patterns entailed in exploiting the environment affect other aspects of culture.

Steward has also indicated that the procedures for the analysis of cultural ecology can be applied more fruitfully in case of any tribal community who depends more on the natural environment for his subsistence economy or survival. As regards the Paliyans perhaps the description of their culture or their customs and traditions as recorded by the administrators and social scientists, may be very useful to identify their traditional culture or to portray their way of life.

During nearly 200 years of colonial rule and more than 50 years of post independence era, Paliyans have been influenced by the external forces, which brings tremendous change in their culture. Therefore, giving an accurate representation of the interaction of the environment with their technology or culture is difficult. Yet an attempt is made here to present the interaction between the previously existing undisturbed environment and the traditional technology and culture as far as possible. By such analysis, subsequent impinging forces of modernization and their impact can be identified and analysed.

## Inter-relationship of productive technology and environment

As described earlier, the Paliyans live either at foot of the hills near to the forest or at altitudes between 2000 ft. to 4000 ft. There are plenty of diascoria tubers and yams which were the staple food of the Paliyans. The Paliyans prepare digging sticks with the billhook, which they purchase from outsiders. In a stony terrain, the digging stick of various lengths can be used as per the location of the tubers and can be brought out with least damage. Such tubers grow automatically or with least labour yet the yields never fail as the failure of crops in drought situation. The fishing practice by 'bunding' and poisoning the shallow fountain is also equally interesting in a rain shadow area. The method of hunting by driving the game animals to a precipitous edge by a group of fellow men and dogs or killing the animals and fowls with the help of device resting loosely on a pole is practised. This destroys the game animals or birds which are smashed with the slightest disturbance to the supporting pole by the game animals. In earlier days, the big animals were made to fall into a pit which was covered with the soil and grass as a result of which they were getting imprisoned. The Paliyans now a days do not hunt big animals but confine themselves to hunting small animals like rabbits or birds with the help of snares or traps. (They might have confined their hunting practices to small animals perhaps due to forest acts). At present, occasionally guns are borrowed and used for hunting big animal stealithily. The environment and culture interact on each other and form into a continuum.

## Analysis of the behaviour patterns involved in the exploitation of a particular area by means of particular technology

The Paliyans while digging roots and tubers go either individually or in small groups. While collecting roots tubers or other food items a couple or a domestic labour group works together and prepare their digging sticks. The hunting or fishing is also carried individually or in group. With the help of their clothes they some times collect fishes. Large scale fishing is done by poisoning the streams. Hunting of animals and birds are done mostly with the help of traps or goading them to precipitous hills. Thus by and large most of the exploitation of the surrounding area is carried individually or by the domestic labour group. Large scale hunting is carried by most of the community members. Some times women work together while collecting mushroom or green vegetables etc. While collecting honey the affinal kins specially brother-in law holds the rope or creeper ladder. Thus specific kin member or husband and wife together play specific roles in exploiting the environment.

## To what extent the behaviour patterns entailed in exploiting the environment affects other aspects of culture

The exploitation of environment is also associated with other aspects of their culture. After the end of the harvest the whole settlement or at least the representatives of the settlement perform the annual ritual of 'Pooma Devi'. On this occasion, all the Paliyans in the hamlet perform the ritual of worshipping 'Pooma Devi' before eating tubers. The rituals are performed before collecting honey or before starting the hunting operations. At present the exploitation of environment by Paliyans and its relationship with leadership pattern and political organization are somewhat disturbed and can not be discussed in detail as the political scenario has been completely distorted by various forces. It requires much more detailed and intensive investigation.

## Summary

Discussing briefly the theory or concept of cultural ecology an attempt has been made to discuss the nature of the habitat of the Paliyans. Through the description of the hills and plateau of Western Ghats, the distribution of perineal rivers and streams and the fountains that dries up in summer, the soil pattern at different altitudes, six vegetational zones as per Puri, and the types of forest in each zone have been described as the environment of the Paliyan habitat. The types of forest that grows in different zones have also been discussed briefly.

In a group interview the informants informed about their knowledge of their environment, how they identify particular place and designate each on the basis of its peculiar features were discussed. In each area of the forest there is a particular deity thus the forest has more shrines and the forest is a more congenial place with familiar localities and spirits. The biotic features, the names of different plants and animals and their use have also been briefly described along with their classifications. Next a section is devoted to ethnomedicine. Due to shortage of space the author could not discuss shamanism elaborately.

The environment and its relationship with productive technology are discussed for a clear understanding of the ecology of the Paliyans.

Finally social structure of the Paliyans is presented by discussing marriage and family, kinship relationship and the village community and the relationship of the Paliyans with other castes and communities.

...

# 3

# The Web of Paliyan Economy

The economic life of Paliyans has undergone a sea change in the late 20th century. In the beginning, most of the Paliyans solely depended on food gathering and hunting activities for their survival. In the extract of Madura Manual 1891 it is mentioned that "Roaming over the hills they satisfy the hunger with such roots and fruits as they can find by search and occasionally a little of honey. They carefully shun the society of civilized men except upon the offer of a piece of tobacco or a strip of cloth, for both of which they show great natural fondness." This fact has been changing gradually. And now after overcoming the stage of barter economy they have entered into the monetized economy. In most part of the Paliyan habitations, it is found that the latter plays an important role in their life at present. In this chapter, the discussion is focused on the nature of present economic activities followed by the Paliyans and their classification (vide chart No. 2).

The economic activity of the Paliyans can be classified as follows:-

Occupation of Paliyans

(A) Gleaning

- (i) Hunting and gathering
- (ii) Collection of minor forest produce (MFP)

(B) Labour

- (i) Work in Plantations or agricultural labour and labour under forest contrtactors and others

(C) Agriculture

- (i) Paliyans as cultivator

(D) Service

- (i) Service holders under government and non governmental agencies
- (ii) Miscellaneous

**CHART NO. – II**

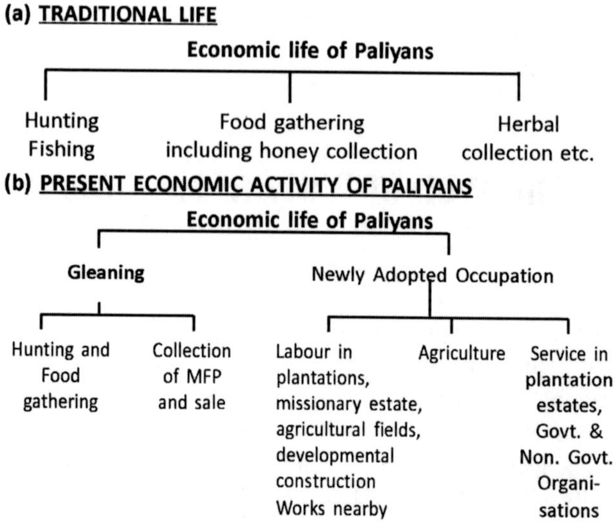


**Note:** Collection of M.F.P. and labour are two prominent activities at present.

As mentioned above gleaning occupations can be broadly classified into two categories:

**Gleaning Economy:** Initial descriptions depict the Paliyans mainly as food gatherers and hunters. They exploit their natural environment by hunting, fishing and food gathering.

In addition, they collect M.F.P (Minor Forest Produce) along with their food gathering economy. Collection of minor forest produce is a very old practice and is presumed to be a continuation from 'Sangam' period. As forest dweller, they eke out their living by adopting these two types of traditional occupations.

Due to culture contact situation they adopt various, other occupations which we designate as "newly adopted occupations". Such occupations are enumerated in the B category. The B category mainly constitutes various types of labour as their principal occupation. These are agricultural labour under individuals and estates, domestic labour and labour under forest contractors etc. Occasionally they also work in the dams and other projects and construction works, roads. Paliyans have contact with outsiders especially with plains people since long.

Though traditionally they pursue food gathering and collection of M.F.P. they are also adopting to the economic life of surrounding environment.

Unlike other tribes the Paliyans do not manufacture any artifacts excepting digging sticks. The use of traps out of iron wire for catching rabbits etc. is a new development learnt from the caste prople. The other Paliyan occupation consists of service category, and includes working in formal and informal sectors. As educational level is generally elementary, some Paliyans have started working as peons, cook, firewatchers etc. in Govt. and Non-Govt. organizations.

It is interesting to note the transition of a food gatherer of yester years being transformed to one having multifaceted economic pursuits in the face of stiff competition from the plains people belonging to larger Tamil society. This also shows the bond between the Paliyans and other community members in their economic sphere.

The nature of Paliyan economic life varies from place to place based on their habitat and other factors associated with the prevailing ecological conditions. One distinguishing feature of their habitat is that it is situated usually on the fringe of forest areas, and their service is highly required in association with the forest environment. They render valuable services and are indispensable for those who have to gain anything from the neighbouring forest areas. Though forest economy was their only resource in the past it is now changing rapidly with the changing new economic opportunities with inundation of forest areas by the caste groups and advanced communities, new avenues are opening for the Paliyans. In the year 1846, a French coffee planter started coffee plantation in the lower Palani hills. Slowly, others followed the pattern and now most of the pockets of lower Palani hills have coffee plantations. This involved cleaning the forest areas in which Paliyans were also involved (Mohanty 1995). So the plantation owners consistently allure the Paliyans and persuade them to work in their plantations and provide shelter. Though this process must have started some 150 years back the seminomadic life style still remains a major obstacle for estate owners and landlords. During my field work in Sirumalai hills, where the plantation had started before the lower Palani hills, a Chakiliyan man mentioned that the Paliyans in their hamlet have not yet given up their nomadic habit despite persuasion. The call of the forest is so strong that they are reluctant to accept any innovation to which they are not familiar.

The present chapter deals with the economic scenario of both the sample villages (details of which is discussed in the previous chapter). Apart from studying details about these two villages, adjoining villages have also been visited to get an idea about the surrounding area. In the beginning an attempt has been made to describe the settlement pattern of both the sample villages.

## SETTLEMENT PATTERN

### Shenbagathope

On approaching the village Shenbagathope from the near by town, by the only motorable road, one would reach at the first settlement. This is the last stop for the town bus operating twice daily to this area. Located on the foothill of a small hillock, this hamlet is sandwitched between private orchards/estates on one side and the huge orchards of the Andal temple located at Srivilliputhur in the nearby town. The temple is the emblem of the Government of Tamil Nadu. The Paliyan houses are built on either side of the road. Here the houses were built by the Government. First set of houses were built in the early sixties. A school was also built along with it. The houses are in a dilapidated condition. Further, second set of houses were constructed in two phases in the eighties. The hamlet has a cooperative society, two tea-stalls and a water tank for supplying portable water to the hamlet. For the piligrims to the Vaishnavite temple, this hamlet is the nearest place to park their vehicles to climb up the temple, as the road beyond this point is not motorable. Though the hamlet is meant for Paliyans only, the tea stall owners and a few others belong to plains area. Their interest lies in trading with the Paliyans, various commodities including minor forest produce and sometimes conspicuous collection of forest goods. Beyond the hamlet, upto the edge of the forest, numerous estates are there, where guava, lemon, mangoes etc are grown. In the last of these estates, bordering the forest, lies the second Paliyan hamlet (locally known as '*Medu*' which forms part of Shenbagathopu). The set up is beneficial to both estate, owners and Paliyans. The presence of Paliyans in the estate help the owners in retaining minimal maintenance staff and protection from the wild animals. Paliyans who want to avoid the police, forest officials and other Government people prefer this place. Again this place also helps the Paliyans to collect the forest produce with ease compared to their counterpart who stay at a distance in the first hamlet. Apart from these two locations few individual Paliyan nuclear families stay in the groves / farms of Tamilians and take care of the gorves / farms. They are paid for their service as mutually agreed. Though there are three settlements in Shenbagathope, Paliyans also stay in the forest encampments when they go to the deep forest. During the field work, researcher could not avail the opportunity to witness these temporary encampments. This was due to the fact that the forest, in this part, was declared to be sanctuary for the flying squirrel, and thus collection of minor forest produce was also banned. The D.F. O incharge of the sanctuary was a liberal person and allowed the Paliyans to collect the M.F.P. from this

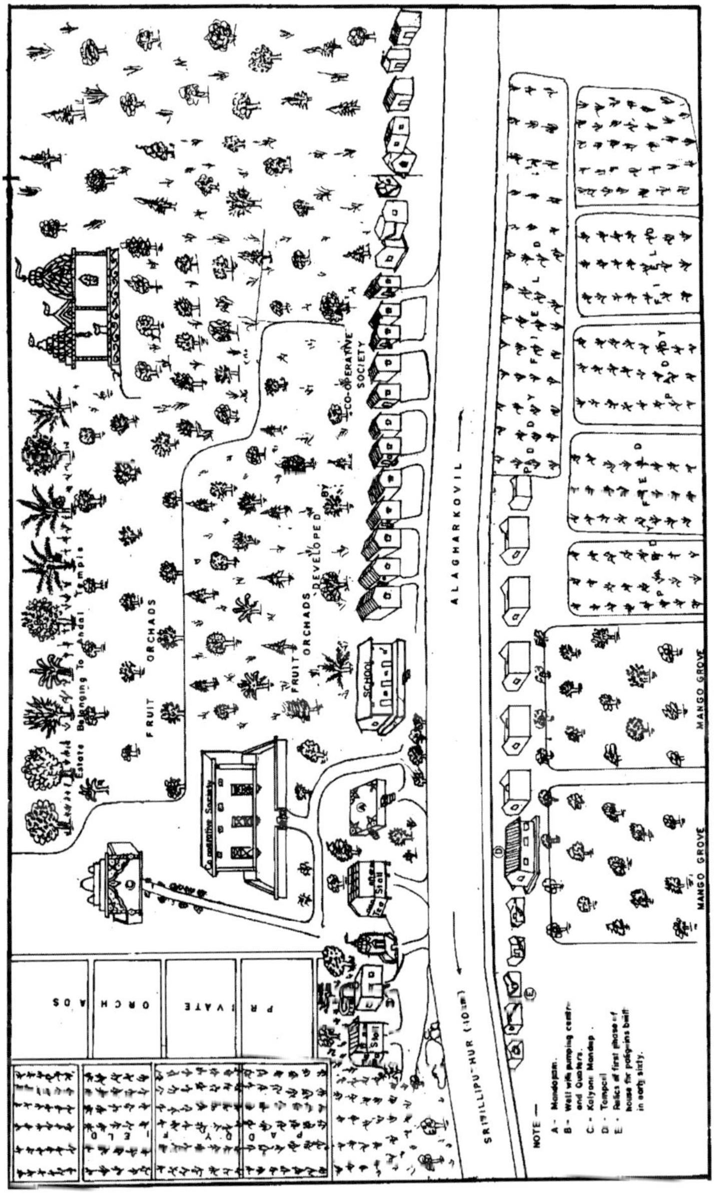
Estate Belonging To Andal Temple
FRUIT ORCHADS
FRUIT ORCHADS DEVELOPED BY CO-OPERATIVE SOCIETY
SCHOOL
Stall
ALAGHARKOVIL
PADDY FIELD
MANGO GROVE
PRIVATE ORCHADS
NOTE —
A - Mandapam.
B - Well with pumping centre and Quarters.
C - Kalyana Mandap.
E - Relics of first phase of house for paliyans built in early sixty.

sanctuary area. If Paliyans have to collect certain items such as resin, lac etc they have to go deep inside the forest as the resin producing trees exist on top of the mountains. Collection of resin involves cutting the bark of the plant and collecting the latex which gets solidified in a slow process. Since Paliyans are getting plenty of minor forest produce near by, they are not very serious for the resin collection. Apart from the availability, the collection of monor forest produce by the Paliyans is based upon the demand of the purchaser. Again, it is only the minor forest produce agent, who has been operating among these people. This man has no legal sanction to buy from these areas. He had engaged a tribal youth who had been earlier the spokesman for these people in interaction, with the outside agents.

In none of these settlements one depends fully on hunting and food gathering for his survival. Hunting is always presented as an act of the past. They reported about fishing activities which are carried occasionally. The main occupation of Paliyans in all the three hamlets is collecting the forest commodities. An interesting feature of the Paliyans here is that it is not possible to identify all the Paliyans as living permanently with specific hamlets. On the contrary, few Paliyans, for various reasons, are associated with the hamlet of their own. Another newly emerging job opportunity for Paliyans in these hamlets is newly created posts of forest watchers created by the forest department.

The first hamlet, as mentioned earlier, has houses built in three phases. First phase houses are completely damaged. The second phase houses are occupied while the third phase houses are roofless and lie deserted. Paliyans live in the second phase houses only.

Behind the houses, on one side, lies the estate of Andala temple (a Vaishnavite temple) spreading over hundreds of acres. Here, there are various types of trees such as mangoes, jackfruit, sandalwood, tamarind, coconut etc. The temple controlling authorities does the collection and sale of these products by auction. Some Paliyans, in every mango season get the scope to work here for watchman's job as well as for other activities. Paliyans also get employed in plucking the fruits. In 1994 summer season, Perumal was working for a private mango grove owner, Perumal received Rs. 40/- a day and was offered food along with some drinks. The rate was higher than the scheduled government rate of a casual labourer. But, the job involved climbing the delicate branches, plucking the fruit, carrying them for loading and unloading. The money received was lower than the wages paid to other Tamil workers for the same job. Paliyans of this hamlet are also engaged as agricultural labourers in nearby fields.

In the second settlement (*Medu*) the houses are of low height with the roof made up of coconut leaves. These are procured from the near by towns by the contractors or estate owners. The size of the houses varies with the number of members residing in it. Out side the houses piled up gunny bags are stored filled with tamarind, 'Kallupattai' etc. As most part of the field study was carried in summer and winter it was seen that, in either case, there was a fire outside the house and Paliyans preferred to sleep in the open around the fire. On some occasions they danced in the center of the hamlet in the moonlight.

A group of Paliyans stay in the estates nearby and are engaged in different odd jobs relating to these estates or groves. Apart from that, they occasionally stay outside such settlements, when they go deep into the forest. There they stay in some caves or rock shelters as per their convenience.

Most of the families have a radio set which acts as the major source of entertainment in their settlement and sometimes at their work site too. All the family members are having minimal utensil as distributed by the Soco (Ngo).

**Table No. 3.1 – Sex-wise Age Group of the Paliyans in Shenbagathope & Palamalai**

| Age group | Shenbagathope | | | Palamalai | | |
|---|---|---|---|---|---|---|
| | M | F | T | M | F | T |
| 0-5 | - | 1 | 1 | 7 | 2 | 9 |
| 6-10 | 3 | 5 | 8 | 6 | 12 | 18 |
| 11-15 | 1 | 5 | 6 | 7 | 7 | 14 |
| 16-20 | 3 | 2 | 5 | 5 | 4 | 9 |
| 21-25 | 3 | 3 | 6 | 2 | 3 | 5 |
| 26-30 | 4 | 5 | 9 | 4 | 6 | 10 |
| 31-35 | 4 | 1 | 5 | 5 | 6 | 11 |
| 36-40 | 2 | 5 | 7 | 4 | 4 | 8 |
| 41-45 | 3 | 1 | 4 | 5 | 3 | 8 |
| 46-50 | 2 | - | 2 | 1 | - | 1 |
| 51-55 | 2 | 1 | 3 | - | - | - |
| 56-60 | | 2 | 2 | - | 2 | 2 |
| 61 & above | 3 | 1 | 4 | 5 | 3 | 8 |
| Total | 31 | 34 | 65 | 51 | 50 | 101 |

**Note:** Population study at Shenbagathope includes that of Pillavakal Dam settlement (17 members).

All the Paliyans have an 'aruval' i.e billhook with them. This is a multipurpose instrument and can be used not only for cutting the trees and plants but also hunting. If a Paliyan is proceeding to the jungle with a dog and having a bill-hook then it confirms that the person is proceeding for hunting purpose.

## Palamalai

This sample village in the Palani hills is located in a valley. Records indicate that this has been an old village[1] of Paliyans. The village is about 10 km. away from the main road linking Kodaikanal and Dindigul. While approaching from the junction to the sample village one would come across various private estates growing horticultural products along with coffee. Almost at the end, one would halt at this lustrous valley. The valley is at present under the control of the missionaries. The sample village is famous due to its historical background coupled with the missionary activities, and it has an identity of its own. Here, there are four hamlets, which are takenup as one integrated village. The estate occupies nearly 250 acres of prime land where coffee, jack fruit, orange, black pepper, banana etc are grown. The estate has built up houses with tinned roof for its permanent staff controlled by the estate administration. This forms the first hamlet. Adjoining the first set of houses, towards south west along the road, second hamlet has come up with 22 houses under group housing scheme of the government. The third hamlet is on the opposite side of the group housing scheme towards East of first hamlet, adjoining the estate territory. This hamlet has come up in last 4-5 years. The motive behind this new settlement was the hope aroused by a local N.G.O. They had promised to make efforts to build a group-housing colony with government aid for these people. The fourth hamlet is located at the North end of the estate. Apart from these four hamlets, there are few houses scattered in neighbouring areas of the estate. These scattered houses are built amidst their own agricultural land. There is a church and a primary school inside the estate. The estate hamlet (first hamlet) has a portable water point, which usually provides drinking water. Efforts were also being made to bring portable water to the group housing scheme area (second hamlet). Most of the provisions are purchased from the missionary managed shop, which opens once in a week i.e. on saturday. There are two shops, managed by caste people, who are working in the estate, and having the shops in their own houses. These shops cater to the immediate needs of the residents. Contrary to the sample village in Srivilliputhur, shopowners here have trade as their secondary activity. In fact, in all the four hamlets other ethnic gropus live alongwith the Paliyans. A detailed discussion on these occupants has been made in fifth chapter. The third and fourth settlement, have thatched houses. Usually '*pullu*' grass collected from the

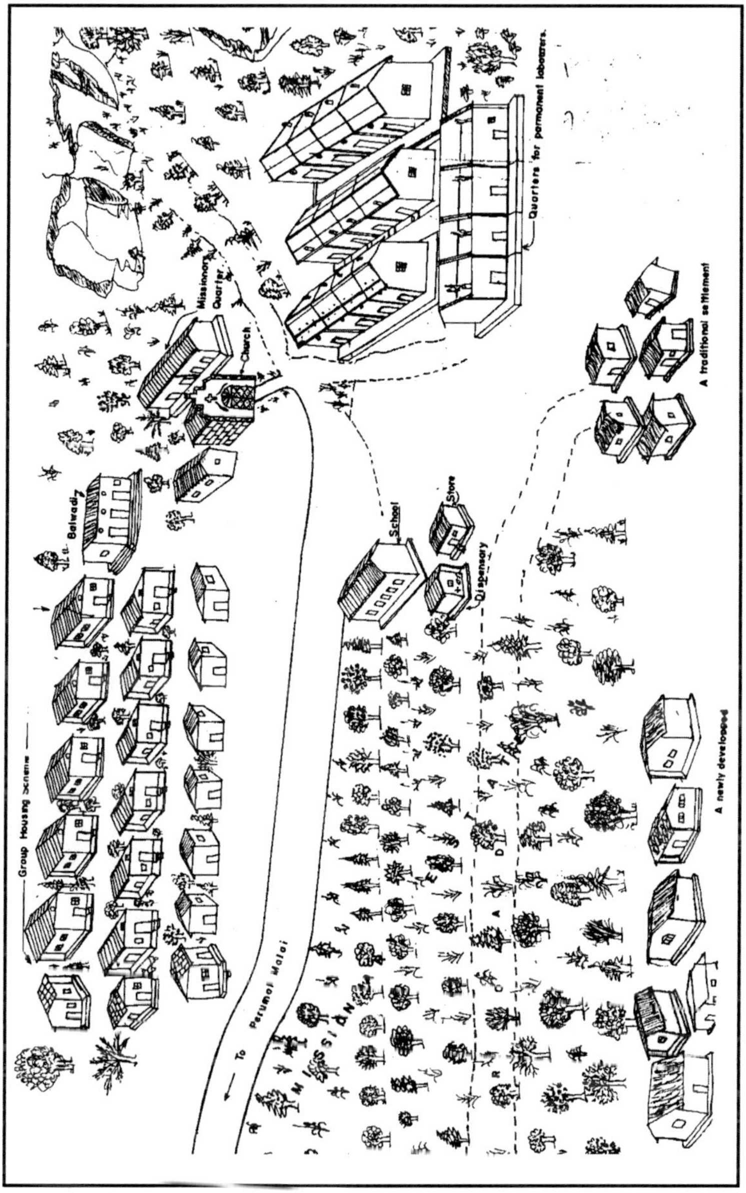
Group Housing Scheme
Balwadi
Missionary Quarter
Church
School
Store
Dispensary
Quarters for permanent labourers
A traditional settlement
A newly developed
To Perumal Malai

nearby area is used in thatching the roof. The individual houses attached to the respective fields have thatched houses. A family spends around Rs. 1200/- annually in thatching the house. Based on the statement of Selvam, head of a comparatively bigger house, the expenditures on different items of thatched roof are as follows:

**Total Expenses**

| | |
|---|---|
| *Pullu* leaves (expenses on collection of leaves) @ Rs. 15x3 days x 3 persons | Rs. 135. 00 |
| Mason with assistant 1 day | Rs. 120. 00 |
| Coconut leaves | Rs. 100. 00 |
| Transportation of coconut leaves from plains | Rs. 100. 00 |
| Labour charge towards bringing poles from '*Penari Maram*' | Rs. 450. 00 |
| **Total:** | **Rs. 1205 .00** |

Of course, Selvam's expenses on house thatching is not a standard for all Paliyans as he is financially better off than most of them. He has two rooms and a cow shed in contrast to single room by the Pallyans in general. He is engaged in stone cutting for which he utilizes other labourers. The above facts give picture of the changing facets in Paliyans life. Most of the people build their houses by themselves and the huts vary but not comparable to that of Selvam either in size or sturdiness. Selvam estimates his house to last atleast for two years.

The individual houses have gardens attached to them. They take proper care of their garden. In these lands, valuable plants such as ginger, coffee, orange, plantain, lemon and various vegetables are grown and sold in the open market. Selvam also grows some of the above plants in his garden. Details about these emerging cultivators is discussed later. There are also others who are staying in the government group houses or in the estate quarters and have their fields adjoining the estate. Though no regular '*patta*' (land records) exist in their name on the lands, they have been encouraged by the missionaries to occupy such land and cultivate it. Many of such properties are jointly owned. There are also landless families possessing only their house. They attribute this state to their migratory habits. Gyanamuthu who is in his seventies narrates how the missionary 'father' used to advise his co-workers to take up a piece of land after referring to some land records. Since he was not regularly working in the missionary controlled estate he missed this opportunity. From the Paliyans, in

Palmalai, it is learnt that Paliyans in Poondi area cultivate garlic, carrots etc. in their fields.

## A. (I) HUNTING AND FOOD GATHERING PRACTICES

Paliyans invariably do not agree openly that they go for hunting even if if is an open secret that they have not given up this favourite habit. (Although they may confess such practice while cross-checked, yet it is very difficult to get any exact information from them on this). The best information on their hunting activities can be gathered from the other community members staying with the Paliyans. Though it is not possible to accept these informations completely yet it can be verified and crosschecked with Paliyan informants who avoid initially telling the details. They avoid talking any thing about this age-old habit as it provides them with valuable wealth. Besides they are afraid of forest officials. The Paliyans who hunt mostly small games adopt very simpe methods of hunting which is an unique feature. It will not be wrong to state that their primitive method of hunting has been a boon rather than a curse in retaining their traditional occupation in a very difficult and challenging situation. The pet dogs of the Paliyans are the prized possessions, as they form the main task force during the hunting expedition. Adaptation of new technology into their hunting pattern has also been noticed. In Palani hills at least in two different hamlets, there were occasions to see, meat obtained through hunting by guns. On enquiry, it was learnt that the animals were hunted with the guns belonging to respective estate owners. Further enquiry revealed that youngsters and good shooters borrowed the gun when they are interested in hunting. The youngsters after becoming intimate to me narrated how on festivities, they joined together to borrow a gun and hunted wild pig/boar (sus scrofa cristatus). After killing it they used to sell the meat for Rs. 20/- a kilo and share the money among themselves. Such organized poaching is done only for larger animals and usually conducted at night. At least on two occasions in a year larger animals were hunted out of which wild pig or boar is more frequent.

Though there is no large estate in Srivillpurthur area as in Palani hills yet Palliyans here (Srivilliputhur) are not spared to have association with persons owing guns. On enquiry with contractors of MFP as to how he has developed contact with Paliyans, he informed that from long past he and his forefathers used to visit these areas for hunting purpose. While doing so they used to take Paliyans with them. The hunting was done through 'Tupaki' i.e gun. Involvement of some Paliyans in an elephant poaching case cannot

**Different types of traps for catching birds and animals**

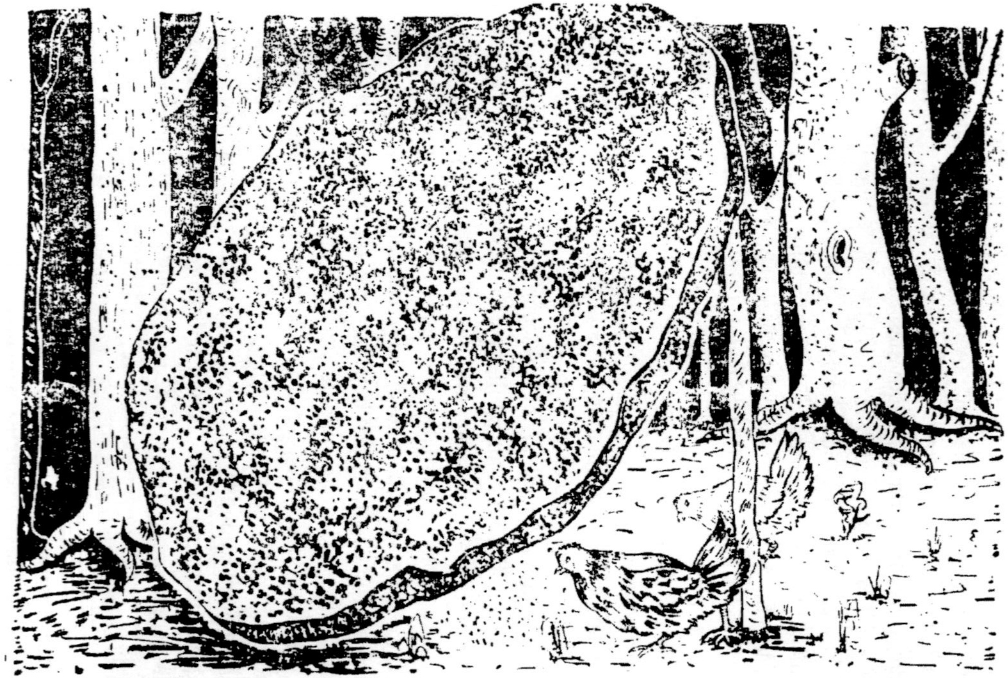

*Source* : Census 1961.

**Idikki – Paliyan way of trapping birds.**

**Rabbit Traps**

be ruled out during my field work. As possessing a gun and hunting in a sanctuary area is prohibited, there is a strong feeling of solidarity in denying about hunting activities. In a hamlet, slightly away from sanctuary area but within a contiguours forest stretch accidentally, I came across a youth arranging a trap. As he could not avoid alling the truth he stated that he had spread the trap at the periphery of the field on the previous day. He was lucky to get two rabbits by the early morning.

In the Palni hills once while interviewing late in the evening, the author observed one of the family members moving suspiciously. On enquiry he revealed that his maternal grand father had hunted two small deer and it was their share that he was holding. On further enquiry with the elderly man, I came to understand that though he was nearing seventy yet he had not given up his habit of hunting. Early morning, he had taken his dog along with the dog of another kin member. The dogs are so trained that they chase any wild animal they could spot. Once they corner the animal it is easy for the Paliyan to kill the prey. The instrument they carry is the '*aruval*'(bill-hook). Paliyans hunt various types of deer. '*Milla*'or '*Melaa*' is obtained through shooting. But hunting '*Mellaa*' is not as frequent as that of wild pig. Once they shared a '*Milla*' (*cervus unicolour niger*) shot by a private estate owner while the animal had come to a saline spot for eating some saline earth. Such large animals are seldom hunted. (Gardner – 1993) Other varieties of deer that they usually hunt are '*Kellemaan*'. (*Muntiacas Muntjac*) '*Pulli maan*' (spotted deer, axis axis) etc. Some of these smaller ones are hunted with the help of dogs as mentioned. Atleast three varieties of squirrels are hunted by the Paliyans. But the Indian giant squirrel (*Ratufa Indica*) and grizzled giant squirrel (*Ratufa Macroura*) provide them with substantial meat. These varieties of squirrel prefer to stay in thick forest so that the canopy can help them to move from one tree to another and thus referred to as flying squirrel. The other animals that the Paliyans relish to eat is '*utumpu*'(Monitor Lizard). During my reconnoiter. I had encountered a monitor lizard on my way to a hamlet. This information helped me to develop rapport as all the people around started querrying about the spot where I have noticed the lizard. This animal usually stays in some hollow of trees or in rock crevices from where it is difficult to remove.

The various types of hunting adopted by the Paliyans is worth mentioning due to the involvement of simple technology in it. In one method they trap certain animals with the help of knots or snares. Paliyans who are well versed with the local forest environment can immediately make a knot, if they spot the animals footprints in appropriate place. The animal when chased will come through this knot. The knot is fixed at such a height that it is just enough for the animals neck. The type of animal is guessed from its foot mark. Once the neck enters the knot it gets tightened and the animal is trapped. Then the Paliyan, if he is present there, kills the animal.

Writing on the basis of Thurston's (1909) description of hunting by Paliyan, Nambiar (1961) states that the Paliyans do not have bow and arrow nor do they have any fire arms with them. Further he states that for catching Sambhur or deer the following method is used "A pit is dug along the path usually taken by the animal and it is covered with twigs and earth. When the animal falls into the pit they rush towards it from the hiding place and kill it. The animal is shared by members of the settlement" (Nambiar: 1961). In the same report there is also a method of catching fowls or rabbits which is known as 'Idukki'. In this method a slab of heavy stone is made to stand inclined over a stick. And grains are put below this stone. While eating the grains the stone is slipped off and it falls heavily on the prey, it is crushed under the stone.

A senior expert hunter in a part of Palani hills described a method of hunting which is based on a simple lever system. Here, on the path of the animal, a triangular structure is erected. In the upper side of hypotenuse, a large stone is suspended very delicately. The base of the triangular structure is made above the ground level in such a way that at the slightest jerk the prop falls to the ground and the heavy weight of the slab crushes the animal. When the animal is chased or even, when it is on its normal course, it hits the base and the upper hypotenuse hand falls on the neck. Thus the animal is killed by this method. They usually adopt this method for catching wild pigs. Nambiar (1961) narrates another method by which the Paliyans try to catch *'gunna'* (*Lacettia iguana*) in the tree holes by blowing smoke into the hole in which it lives. It is also reported that the Paliyans catch rats in the tree holes in the month of '*Adi'*.

Catching fish is another hobby of Paliyans. One of the unique feature of catching fish is by poisoning the water of the stream or river. In the Palani hills, in certain pockets near to Paliyan settlement one can find a shrub belonging to the family of 'Papilinecea'. When going for fishing they carry sackful of these leaves and flowers along with them to the stream. They try to block the water with stones so that flow is checked. Then, they crush the leaves of the plant with a stone and put it in the water. After sometimes fishes start floating on the surface. I asked whether the poisoned fish would create any health hazard to them. They replied in negative. They believe that cleaning of the fish thoroughly takes care of this problem.

Nine decades back, Father Dahmen while narrating about the fishing activities has stated that they use the leaves of a creeper for poisoning the water. But in Srivilliputhur area they informed that fishing is done in the forest stream or in other water bodies by using the sari or long cloth that is available with them. During my stay, one of my key informants offered the fish curry. When asked how they have got fish when they had gone for collecting tamarind he narrated that on their return they took bath in a stream and while doing so his sister and grand mother used their sari to catch the fishes.

In the recent research pertaining to hunter gatherer studies it is observed that hunting is the dominant subsistence mode only in the highest latitude, but in the remaining area gathering is the dominant activity (De'vore 1972, Morris 1982). Food obtained from hunting of mammals is as low as 10%. It is impossible to quantify the amount of food obtained by Paliyans from hunting activities due to concealment of facts as mentioned above.

Paliyan's subsistence activities in Palani hills differ from that of Srivilliputhur or the Sirumalai at least in one way i.e., Paliyan hunting and gathering activities, Paliyans in Srivilliputhur and also in other areas where they lead a traditional life style do not assign their timing separately for hunting and gathering activities. They normally wander in small foraging parties, usually consisting of nuclear families. They collect whatever comes across in their way, marketable forest produce, plant food or game animal's etc. This does not imply that they donot pursue hunting specifically on any occasion. On the whole, usually foraging activities still remain a major pursuit of their life and thus they return home with a mixed bag of marketable produce, wild foodstuffs and game animals. As market economy is making an impact on their life style, collection of marketable produce has become important and thus many foraging parties have become specialised in gathering specific commodities. On the contrary the scenario is different in Palani hills. Majority of the Paliyans encountered in the sample village and the peripheral villages were engaged in estates. Thus Paliyans in this part enter into forest with the purpose of hunting. And thus hunting was a preset mission.

## Food Gathering

Apart from hunting, Paliyans are also involved in food gathering practices till today. It is true that the food gathering habit has been reduced but it still remains as an interesting activity for them. They mainly depend on some species of tubers and other things such as fruits, flowers, leaves,

***Paliyans gathering honey***

roots, bulbs, stems, mushrooms etc. In Chapter – II some of the fruits, collected from trees and creepers, items of seeds and greens are mentioned which the Paliyans usually collect from the forest for their consumption. Some Paliyans in certain pockets are expert honey collectors.

## A. (II) COLLECTION OF MINOR FOREST PRODUCE FOR TRADERS

Invariable all the Paliyans are involved in collection of minor forest produce. On behalf of the state, forest department controls all the activities connected with it. The collection of minor forest produce is an important aspect of Paliyans economic system. In between the Paliyan and forest department, come M.F.P. contractors. Like most of the forests in India, here also the M.F.P. collection is done by the contractors holding such right for a specific range. They are selected on the basis of highest bidding in auction of the working circle forest range for a year (i.e. July to June). It is generally referred to as M.F.P. working circle and it covers all the reserved forest and unreserved land of the forest division. The control of the exploitation of both these types of forests is vested with the concerned D.F.O. In the Kodaikanal Forest Division the M.F.Ps. are collected under three groups.

**Table No. 3.2 – Varieties of M.F.P. collected from Kodaikanal Forest Division.**

| Sl. No. | Group – A | Group – B | Group – C |
|---|---|---|---|
| 1. | Avaram bark (Cassia Auriculate) | Tamarind | Deer horn |
| 2. | Beewax | Myrobalam | Sambher |
| 3. | Bothai grass – Lemon Grass etc | Lemmongrass | horns etc. |
| 4. | Dwarf Date leaf (Phoenix Sp.) | Konari bark etc | |
| 5. | Galnut (Terminaiia Chebula) | | |
| 6. | Honey | | |
| 7. | Konnari Bark (Cassia Fishtula) | | |
| 8. | Marking Nut (Semi Carpus Apacardium) | | |
| 9. | Neem Seed (Azadirchata Indica) | | |
| 10. | Nellikai (Phyllanthus emblica) | | |
| 11. | Shekakoy (Acacia Concinna) | | |
| 12. | Soapnut (Sapindus Emarginatus) | | |
| 13. | Stone and tree moss – Pasam (Lichens) | | |
| 14. | Wood Apple (Feronea Elephantum) | | |

The Group – A items are auctioned in lot and given to one contractor.

The Group – B items are individually auctioned by the department.

Forest department directly collect the group – C items through their own personnel.

Unfortunately the tribals in this area do not enjoy any legal right over the collection of these M.F.Ps. So the M.F.P. contractors treat the forest product as their own. They see to it that the Paliyans sell the M.F.P. collected by them to their agents only. In the sample village in Kodaikanal (Palni Hills) as many as three agents were collecting gal nut for the main contractor. These agents took opportunity to make some extra money by cheating the Paliyans. For instance, gal nut is never weighed but it is measured by 14 litre oil tin. All the three agents have increased the length or height of the sheet, so that the tin carries some extra gallnut in it. Interestingly, Paliyans sometimes pass on the produce to the agent of another area as this hamlet is located in the border. This is done for better gain and for good will. The prices the products yield in market is significantly different from what Paliyans get. It is only at the mercy of the contractor that the Paliyans get the price of their collection from the most inaccessible area. No exact information about the actual collection of M.F.P. could be obtained. Rules stipulate submission of monthly accounts of M.F.P. collected and dispatched by the contractors before 10th of every month to the range office. But the accounts submitted are grossly manipulated to conceal real income and therefore unreliable. However, an idea of profits that the middle man makes can be estimated from the following observations. The estimated revenue and expenditure in the minor forest product working circle is as given below (as collected from the forest department).

The Kodaikanal forest division comprising both reserve forest and revenue forest has an area of about 40740.48 hectares. In this vast stretch, various M.F.Ps. as mentioned earlier are collected both by forest department and the contractors.

**Table No. 3.3 Revenue and expenditure pattern in the Kodai canal forest division from collection of M.F.P.**

| Year | Revenue | Expenditure |
|---|---|---|
| 1990 – 91 | Rs. 49000/- | Rs. 4000/- |
| 1991 – 92 | Rs. 14000/- | — |
| 1992 – 93 | Rs. 49000/- | Rs. 4000/- |
| 1993 – 94 | Rs. 14000/- | — |

Collection charges for few items of forest produce paid by the contractors and also the department are as follows.

**Table No. 3.4 – Rates of collection charges paid to tribal's / others for M.F.P.**

| Sl. No. | Name of the Produce | Unit of Quantity | Collection charges (in Rs.) |
|---|---|---|---|
| 1. | Myrobolams | Per Kg. | 00.60 |
| 2. | Honey | Per Kg. | 04.00 |
| 3. | Bee Wax | Per bg. | 02.00 |

Gal nut is collected seasonally for about four months. The collection starts even before it reaches full maturity. The green ones, which are not fully matured are collected for Rs. 10 to Rs. 12 per tin. The tims used for measurement (as mentioned earlier) roughly carries 15kg. of gal nut. The fully matured ones are collected @ Rs. 30/- a tin. In the sample village at Kodaikanal. I enquired the local agents about the quantity of gal nut normally collected from this area. They replied that nearly two truckloads of gal nut leave the village in a single season.

The traders sell these gal nuts in the district headquarter where the material is in great demand for tanning. In the local market, mixed varieties of gal nut are available at 2-3 times the collection prices. Similarly honey fetches at least ten times more than the collection price. But the total revenue collected from the division looks very minimal. It is clear that the contractors and others gain the lion share from the MFP transactions. In this area, as most of the Paliyan's primary duty is labour in the estates and else where, collection of M.F. P. is the secondary activity. Even then they have not yet given up their habit of collecting these minor forest produce. Such collection varies from place to place and in a similar situation, it varies from individual to individual. Galnut is the most significant M.F.P. that is collected in this village. Only two families, out of the 20 Paliyan families identified here have the MFP collection as primary occupation. Mr. Mariasilvai has not taken up this profession out of his choice. He was earlier working in the missionary estate and was content. But due to difference of opinion with the management he had to quit. He is still prepared to return to work in the missionary estate if he is given the opportunity. Since he is not very sturdy and handicapped, the private estate owners offer him job only when none is available and on contract basis. Now that he is out of employment he has focused on forest produce collection. To encourage more collection one of the agent allures the people to go for M.F.P. collection so that if a person supplies above 100 bags of gal nut he would be offered a *dhoti* and so on.

During the collection season he had given a *dhoti* to the highest collector (Mr. Mariasilvai) who had collected around 125 tins. The second person who prefers to have M.F.P. collection as his primary economic activity hails from a distant place and incidentally is the most efficient honey collector in that place. Apart from these two people all others are involved in some way or other in the collection of M.F.P. as their secondary or leisure time activities.

*'Pasam'* (a lichen growth on the trunks and branches of the trees) are collected by Paliyans belonging to nearby hamlets. These are scrapped with a steel knife and then collected in a bag. One kilogram of *Pasam* is given to the trader for Rs. 16/- to Rs. 18- In the first year of field work, the author had observed that many Paliyans were involved in these activities. But in the following year the forest department had banned this activity as they felt that this was harmful for the plants.

Different items are collected in different months. Unfortunately there is no organization to buy these produce except that of honey collection society at Thandikudi. So some times honey is sold to local people instead of going to Thandikudi as it fetches less amount.

In the second sample village in the Kamarajar district, collection of minor forest produce formed the primary source of income for all occupants excepting those who were employed as fire watchers by the forest department. This hamlet was within the area declared to be one of the sanctuaries for flying squirrels and collection of MFP was not allowed. Yet the district forest officer though it best in the light of changing forest rules not to disturb the forest dependency of these people. He allowed them to collect the minor forest produce to meet their needs. This area contains good number of tamarind trees and during the period from May to July Paliyans have a good source of income from its collection. While the work is not very arduous yet the income generated through such activities is high. Normally it is possible for a person to collect half bag of tamarind (unopened) by working for 5-6 hours a day Each kilo gram of tamarind without opening the thick outer cover is valued at the rate of Rs. 5/- per kg. The work is simple. The process starts with the identification of trees having matured fruits. Once this is done usually adult males climb the trees and shake the branches. Women are not prohibited to climb the trees. As the ripened fruits fall down, the women and children collect them. Though the nuclear family forms the basic unit, yet in this economic activity it is found that occasionally two to three nuclear families joined together.

**Case study:** Once, the author had the opportunity of accompanying a small group of Paliyans consisting of a couple, the father of the woman, a boy of around 10-12 years and another person. It was interesting to note the cooperation and competition followed in the activity. Both husband and wife formed a unit and the other two adults formed another. The husband and the other person climbed the tree. As both these people shook the branches, the respective partners i.e. the wife and her father collected up the falling fruits. They also collected the fruits for their partners who were on the trees. While this was going on, the small boy was made to wait for a short period. As she collected some fruits for herself and for her husband, she signaled the boy to join. The boy after collecting for himself with the lady went to the other side where the old man (wife's father) was collecting for himself and the other man on the tree. This continued for nearly four hours during which they covered three huge trees. At the end, it was decided that rice gruel is to be prepared. Immediately the young lady went to the dry river bed and started to dig a hole by pushing aside the sand in her hand. Then slowly she started to collect the water under the sand with her palm. Within half an hour 3-4 litres of water was collected. Subsequently, the idea of preparing the gruel was given up. By then each of them had collected 10-15 kg tamarind fruits. On the way back we met another widower who was returning back from forest with his collection of tamarind. He was the sole earning member in his family but did not want to depend on any one. Though his elder son was with us he went alone. This man never depended on his elder son for any economic support even though he maintained his second son aged 6 years. The man was over 65.

In the month of February, the author had the occasion to witness the collection of 'Kallupattai'. It is the hard scaly covering of a large tree. The trader had fixed the rate as 0.70/- per Kg. Its utility was either not known to the tribals or they did not reveal the truth. The author had collected the weekly collection of Kallupattai by Paliyans as given in table 3.5 below.

The quantity of this item collected by a couple varied from 38 Kg/day to about 54 kg/day. A boy of around 11 years had collected 35 k.g. in two days period. The reason of this difference is related to the ideology of the individuals concerned. Older people, around 50 and above, collected only the quantity needed to meet their minimum requirement. Younger Paliyans have external exposures and better acquaintance to the monetized economy. For instance Mutheya is an efficient collector of M.F.P. So also is the case with Thangaraj who works in the tribal school and also collects the M.F.P. Mutheya and Thangaraj etc. have a desire to earn more. Normally, every one goes for collection of tamarind in day time. In the evening they remain drunk and spend time with family members and neighbours in gossiping and other entertainment activities. On the contrary, Mutheya and his family keep themselves busy in the evening and in the night by removing the cover of the tamarind. Thus, they can sell the tamarind at a higher cost and earn more.

**Table No. 3.5 – Weekly collection of M.F.P. at Shenbagathope**

| Sl. No. | Name | Sex | No. of days worked for k. Pattai | Amount Collected in kg | Age | Amount obtained per day per capita | Cost of the collected amount per day |
|---|---|---|---|---|---|---|---|
| 1 | Chelleya & Tayaman | M F | 5 days | 189kg | 55 52 | 19Kg. | Rs. 13.30 |
| 2 | Vellamal | F | 1 Day | 40 Kg | 30 | 40 Kg | Rs. 28.00 |
| 3 | Veeramal & | F | 7 Days | 372 Kg | 35 | 26 Kg. | |
| | Ponneah | M | | | 40 | 57g | Rs. 18.60 |
| | 4 Baltai | F | 7 Days | 171 Kg. | 30 | 24 Kg. 43g | Rs. 17.12 |
| 5 | Krishnan | M | 2 Days | 35Kg | 11 | 17 Kg. 50g | Rs. 12.25 |

Another economic activity for Paliyan is the collection of herbal medicines for both human use and animals. The involvement of Paliyans in collection, use and selling the plant and plant products is the highest in Srivilliputhur range. Presence of Ayurvedic college and other allied medicine producing centres in Tirunelveli helps in commercialization of this business. An agent from Srivilliputhur regularly visits Shenbagathope village for placing order on various plant and plant products. He has listed about twenty eight species of plants collected by engaging the Paliyans. Some times the Paliyans are mere collectors of these plants and do not know its use. For instance *'sakaraikolli' elai* is obtained with the twigs at the rate of Rs. 3/- per kg. If only the leaves are given it costs Rs. 20/- a kg. Most often the agent comes with a plant and asks the Paliyans to collect the same. Paliyans exhibit ignorance on its application and use. They have the medicine specialists among them. Neighbouring caste villagers sometimes come and consult these persons for their ailments. On enquiry one of the medicine man replied that since his child-hood he had been roaming over the hills. Sometimes, some Ayurvedic specialist came to the area in search of herbs and plants etc. During the interaction with them he had come to learn about different plants and diseases it could cure. Slowly he started giving medicines and became popular. In another hamlet there was a Dhevar man married to a Paliyan woman and spent over 50 years among Paliyans. Practically for all purpose he was a Paliyan. According to him, his knowledge about medicinal plants was from two source. One, his faher adopted the profession of collecting herbal plants. Other, his close association with Paliyans had helped him to acquire this knowledge. Thus he had been able to combine the knowledge of Ayurvedic Medicine and traditional Paliyan folk medicine in treatment of his patients.

## (B) (I) PLANTATION WORKERS OR AGRICULTURAL LABOURERS

The Palani hills where only Paliyans and Puliyans inhabited, was slowly populated by other castes who migrated from the plains. Some four or five hundred years ago, a caste of Vellalars from Coimbatore migrated into the Palani hills. Subsequently, during the 17th and 18th century some villagers from the plains migrated to the Palani hills to avoid the exploitationof the Vijayanagar Kings, Mahrattas and Tipu Sultan. Its population was later added to by evacuees by the cholera stricken Shembaganur (Bruce De jong 1971).

The infiltration of farmers from plains had their impact on the nomadic Paliyans. During colonial period when large scale migration of plains people took place into Palani hills, they engaged the roving Paliyans to work as labourers.

Gradually they employed Paliyans in grazing cattle and as unskilled agrestic labourers in the fields (of those peasant castes).

The introduction of the reserve forest Act in 1881 initiated problems for Paliyans. On the other hand agricultural activities in the surrounding areas and prolification of plantation or estates provided new opportunity to work as agricultural labourers.

In the case of Palani hills most of the Paliyans inhabiting the area are wage labourers. They work in the plantation for fixed hours everyday and all their earnings are paid at the end of the week. The sample village taken up for study is under missionary control, but the pattern of labour varied. Some are employed as permanent employees and this helps them to receive the bonus and other benefits from the plantation owners. The others are temporary workers and are deprived of various service benefits.

As, employing both casual and permanent labourers are commercially beneficial, most of the plantation owners keep only a few trusted labourers, as permanent employees. When the number of permanent workers increase, the estate owners have to spend more money on fringe benefits under the labour law. Another factor that troubled the plantation owners was the health care measures adopted for permanent employees, as the expenses are borne by them. In this missionary maintained estate the workers assemble at 6.am. Those who have been baptized go for a prayer in the church preceding their daily work. They work in the estate spanning over 250 acres in the rich valley, for 6 to 6 ½ hours. The estate has orange, jack fruit, coffee, black pepper plantation and a small patch of cardamom plantations.

The labourers include men, women and children. Out of the total population taken up for the present study nearly 62% are workers. Among them nearly 50% are women. Although most of the workers are in the age group of 15-55 years yet there are instances where children in the 'dependency category' i.e. below 15 years take up jobs. In at least three instances children were found to be working as servant/helper in distant places. While the parents of two children are temporary workers, the son of a permanent employee has runaway to work in the neighbouring state Kerala. Sometimes, some children take up temporary jobs with contractors who employ them in plucking the fruits like orange etc. The degree to which children in the dependency group take up job is significantly less. Many of them are enrolled at the primary school where they are also provided with noon meal by government.

Since children grow up inside and around the estate, they are exposed to the nature of works performed in it. It was observed that where the Paliyans had their own land, involvement of children in agricultural activities was greater.

So the labour in Palani area can be categorised as follows:

(1) Contract Labour

(2) Workers in estates.

**Contract Labour:** In Palani hills, the estate owners and individual land owners employ contract labour. This can be of two types. In one, they prefix the rate for doing a work. In one of my observation a Dhevar man had negotiated with a Paliyan youth to till the soil, clean up the pebbles and trim the trees. The wage was fixed for Rs. 250/-. The plot was small to assess the labour cost. Unlike other plots, this plot had lot of stones in it and tilling this plot required hard work. After even seven days the work was still incomplete and required another 3 days of labour. The amount worked out to be Rs. 25/-a day. The man worked from 7 am to 6 pm. With a break for lunch i.e. around 10 hours/day. The owner was constantly encouraging about the strength and capability of this person by saying "This man is a serious worker. He is so capable that he can complete this work within 5 days. Out of laziness he has been postponing his work". The owner thile appreciating this scheme mentioned that the arrangement was advantageous to the workers as they had the convenience of working according to their own schedule and will. Another kind of contract labour usually prevalent is allotting a defined area for a given job for a fixed price. For example a piece of land measuring 10 cubit by 10 cubit, (appx. 15'x15') locally known as '*sadara*', is taken. This is tilled for Rs. 25/- as a standard rate. Accordingly necessary calculations is made for doing a job. Another ardous work done by a Paliyan here is stone cutting. He has learnt this new trait with his contact with outsiders. He is proficient in breaking open large stones with gelatin sticks and then grating it in small bits.

**Working in the estate:** Details about working pattern in the estate has been discusswed earlier. From the work force only the regular workers are benefitted from getting the permanent job. As they moved from one place to another they lose the opportunity of getting money regularly. But daily wages were uniform for both temporary and permanent workers in the missionary estates. All the adult males received Rs. 18/- per head per day and all the females received Rs. 15/- per head per day. Youths in the age group of 15-18 years received Rs. 15/- per head per day equivalent to the wage of women work force. The only octogenarian Maria Nandam received Rs. 15 per day. When enquired why the payment was low, the old man replied that since he was going to forest to graze his cattle, the missionary father asked him to take his cattle along with missionary cattle, so that it would be convenient. Therefore, he was doing the job at a lower rate.

## (C) PALIYAN AS CULTIVATOR

The Paliyan is intimately connected with the forest and their occupation is by and large dependent on the forest resources. Though these people live in the rural areas they have not yet adopted to the life of land owning agriculturist. Probable the success of Paliyan as a cultivator will be warmly welcomed by forest management lovers and those who feel tribals come as a barrier to forest development. Though this process has not been a real success yet the rudimentary success of the same is note worthy in certain cases.

In Palamalai the cultivation has been a success due to the involvement of the Paliyans. This is the village where missionaries have set up their estate almost a century back. In addition to 205 acres of 'patta' land they possess some more lease lands from the government is included. Others lands are gradually encroached & acquired by the estate on various occasions. In one such incident a bit of land measuring 12 acres was proposed to be merged with the estate. But it is reported that the application was rejected as the area was shown to be an original habitation of Paliyans. Then the missionary father (estate manager) thought of an alternative. It was reported that he saw to it that this land could go to some of the reliable workers in the estate. While this was going on some of the plains people who had been settling down around the estate started to occupy the revenue land.

**Table No. 3.6 – Land owned by the Paliyans.**

| Name of the village | Land Owned by individual family | | | | |
|---|---|---|---|---|---|
| | Upto 1 acre | 2 acres | 3 acres | 4 acres | More than 4 acres |
| Shenbaga Thope | — | — | — | — | — |
| Palamalal | — | — | 3 | 2 | Four families jointly share a piece of 12 acre land. |

While doing the field work it was very impressive to find that some of them have been reaping the benefits from cultivation. One of the informants who is around 80 years old has four sons and a daughter. He has around 10 to 12 acres of land under his possession. Though he is not taking any benefit from it, he has left it to his children to get the benefit from the same. The land has gallnut, coffee, orange, plantain plants etc. Though the property is not divided yet there is some kind of understanding between the sons. As the coffee plants are planted by the eldest son and taken care by his family, all other family members take only the amount of coffee that is required for their domestic consumption. The excess amount is sold by the eldest son. Similar is the case for orange. Galnut grows in this part by itself.

There are over a dozen of this plant in their land. The fruits from these trees are taken by the brothers on rotation. A part from that, one brother grows a special kind of plantain very popular to these hills which can fetch a price of Rs. 90/- per 100 fruits. Some land is alos cultivated for beans and ginger by each of the brothers. They have now learnt about the agricultural practices and tells how rotation of crop is advantageous to the farmer. Ginger when cultivated in a piece of new land gives a yield of around 4 to '10 times the quantum of seedlings sown in the field. But since that cannot be done year after year they plant beans in the fallow period. Beans is also attractive for them as it can yield within a period of 3-4 months. They cultivate three varieties of beans in their land. While cultivating the beans they take all precautionary steps. Plantain is also an important crop due to the attractive price it yields. Unfortunately the particular species are suffering from some fungal attack and hence involves lot of risk. The remedy for this disease is yet to come. So Paliyans are not able to plant them. Regular care of horticultural plants such as orange, lemon etc. are not very encouraging among many though some have started getting yield from it.

In another hamlet the Paliyans got 'bhoodan' land from ASSIFA (a voluntary organization). The land is dry and though they have been provided with group houses adjoining to their land, they are mostly depending on forest economy. Though they cultivate gingily they do not seem to be very much impressed with the yield.

## (D) PALIYANS WORKING UNDER GOVERNMENT AND PRIVATE ORGANIZATIONS

There are not many Paliyans under government or private agencies, yet there are lot of opportunities for them. In one of the G.O. Ms. No. 520 Environment and Forest Department dated the twentieth August 1991, the Forest Department has taken a stand to employ tribals residing in tribal settlement within jurisdiction of certain sanctuary areas as forest watchers relaxing relevant rules relating to age, physical and educational qualifications. This has greatly helped some of the paliyans as the hob did not involve any regular work. They were supposed to keep an eye on the movement of the people and report anything that they felt to be illegal. Apart from that they were supposed to join other forest department staffs whenever there was any raid or such other works. While staying in one sample village the forest watchers went on such raid on two occasions. It was observed that though their salary was low yet the job was accepted with pleasure.

In the missionary managed estate the village sub-post office is headed by a Paliyan man who is literate up to primary level. His education has helped him to get the job. As voluntary organizations have come up in most of the hamlets a new type of job is emerging among the literate Paliyan youths. In the initial stages the organizers were hiring other caste people in the post of teacher etc. Now they have started employing the Paliyans. This, they feel will help them in getting closer to the group and coordinating the work in a smoother way. The pay here also is very less and vary from Rs. 250 /- to Rs. 600/- p.m.

**Table No. 3.7 – Number of service holders in two villages**

| Name of the village | Number of persons holding Govt. service | Service in Estate | | N.G.O 's Service |
|---|---|---|---|---|
| | | Regular | Temporary | |
| Shenbaga-thope | One (cook in school) Four (fire watcher) | — | Seasonal labourers | One |
| Palamalai | One (Post Master) | 9 | 19 | Four |

While the occupational description given above highlights the different facets of Palians way of life, the economic picture on expenditure and consumption confirms their inclination or adoption to modern monetized economy.

**Table showing consumption pattern and Income**

| PALAMALAI | | | | SHENBAGATHOPE | | | |
|---|---|---|---|---|---|---|---|
| No. of Family members | 6 | | 2 | | 3 | | 4 |
| Weekly Average Consumption Per head | 48.60 | | 63.90 | | 75.75 | | 51.85 |
| Weekly Average Income per family (Approx.) from all sources | 390.00 | | 210.00 | | 400.00 | | 390.00 |

**Table No. 3.8 Table showing provision List for a week in both Palamalai and Shenbagathope**

| | PALAMALAI | | | | SHENBAGATHOPE | | | |
|---|---|---|---|---|---|---|---|---|
| PROVISION | FAMILY – 1 | PRICE (In Rs.) | FAMILY – 2 | PRICE | FAMILY – 1 | PRICE (In Rs.) | FAMILY – 2 | PRICE (In Rs |
| RICE | 15kg | 90.00 | 9kg | 54.oo | 15kg | 97.50 | 20kg | 168.00 |
| Dal | 3/4kg | 18.00 | 1/2kg | 12.00 | --- | --- | 400gm. | 10.40 |
| | | | | | | | | |
| Cuminseed | 50gm | 2.00 | 25gm | 1.00 | Masala or mixed powdered condiments | | | |
| Corlander seed | 150gm | 3.00 | 50gm | 1.00 | | | | |
| Turmeric | 50gm | | | | | | 100gm | 2.00 |
| Chilles | 100gm | 2.00 | 100gm | 2.00 | | --- | --- | --- |
| | | | | | | | | |
| Tamarind | 200gm | 1.60 | 100gm | 0.80 | 100gm | 3.50 | 100gm | 3.50 |
| | | | | | | | | |
| Palmyara sugur | 2kg | 30.00 | 1/2kg | 7.50 | --- | --- | 1/2kg | 7.50 |
| Salt | 1/2kg | 0.50 | 1/2kg | 0.50 | 1/2kg | 0.50 | 1/2kg | 0.50 |
| | | | | | | | | |
| Vegetable | 3 1/2kg | 37.50 | --- | --- | 1kg | 5.00 | 2kg | 10.00 |
| Onion | 200gm | 1.00 | 200gm | 1.00 | --- | --- | --- | --- |
| | | | | | | | | |
| Gingily Oil | 200gm | 7.00 | 100gm | 3.50 | 200gm | 7.00 | --- | --- |
| Coconut Oil | 100gm | 5.00 | 50gm | 2.50 | 200gm | 10.00 | --- | --- |
| | | | | | | | | |
| Drinks | | 6.00 | --- | 30.00 | --- | 50.00 | --- | --- |
| Bidi | Four Packets | 8.00 | 2 Packets | 4.00 | 5Packets | 10.00 | --- | --- |
| Match Box | Four Packets | 2.00 | 2 Packets | 1.00 | 3 Packets | 1.50 | 3 Packets | 1.50 |
| Soap | Four | | 2 Packets | 7.00 | 2 | 11.75 | 2 | 4.00 |
| Chit fund | Saving | | | | | | | |
| Misc (Utensil) | | | | | | 30.00 | | |
| **Total** | | **267.60** | | **127.80** | | **226.75** | | **207.40** |

**Table No. 3.9 – Major expenses in a Annual year in both Palamalai and Shenbagathope**

| Items | Palamalai | | Shenbagathope | |
|---|---|---|---|---|
| | Family – I | Family – II | Family – I | Family - II |
| Cloths | 1580 | 727 | 535 | 420 |
| Medicine | | 15.0 | 650 | 40 |
| Utensils & Funiture | | | 63 | Received a set freely |
| Ornaments | 1300 | | 1175 | ---- |
| Transport & Communication | 700 | 600 | 200 | 300 |
| Gift giving | 156 | 147 | 131 | 110 |
| Purchase of Raido, Tape | | | | |
| Cassettes, Batteries | 1780 | | 400 | 300 |

**Analysis of the Table:** Table 3.8 indicate weekly consumption in two villages i.e Palamalai and Shenbagathope. Table 3.9 gives an account of major items purchased in a year. The families survey belonged to extreme ends i.e at lcast one economically well off / better off family and the other, a poor one. The weekly expenditure on food at Palamalai and at Shenbagathope indicate a marginal variation. At Palamalai it was between Rs. 48. 60 to 63.90 per head and at Shenbagathope Rs. 75.75 to Rs. 51. 85 per head per week. The expenditure is higher at Shenbagathope than Palamalai owing to the fact that Paliyans in the former habitat earned there living through sale of M.F.P . However sales were not constant and varied according to the season, the capacity of the individual collecting M.F.P. price fixed by the contractor and will of the forest officials. On the other hand the income of Paliyans at Palamalai is observed to be a fixed regular amount as they were regular daily wage earners in the estate. It was observed that, when Paliyans at Shenbagathope received the money they spent whimsically sometimes spending a major portion on wines. But Paliyans at Palamalai plan their expenditure rather carefully owning to their regular income.

Table 3.9 reveals that affluent Paliyans in both the settlements spent more money on clothes and ornaments. Apart from this, they also spend highly on entertainment items like tapes, radio etc. This indicates inclination of present day Paliyans towards modern life compared to yesteryears. One notable observation is that people at both the village have expenditures on transportation indicating travel to places other than their habitats and gifts to others as a gesture of love and affection.

It was observed that the family with higher income at Palamalai had earnings from coffee, ginger and other harvest and they spend more on ornament and entertainment. The income of Paliyans at Shenbagathope has increased recently due to intervention of government officials and N.G.O.'s. Exploitation by contractors during M.F.P. collection has decreased than before. Earlier the contractors took away most of the sale proceeds, giving little to Paliyans. It is worth noting that prior to declaration of adjacent forest as sanctuary at Shenbagathope, contractors were leased for collection of M.F.P. and Paliyans used to get a paltry sum from the collection. As a consequence of declaration of sanctuary collection of M.F.P by contractor was officialy banned. But forest officials did allow Paliyans to collect some M.F.P. The latter sold their collection in the market / agents. This resulted in higher income for Paliyan at Shenbagathope than before.

The above table (3.8 & 3.9) is prepared from the data collected through interview and participant observation. In the case of Palamalai collecting data was more authentic due to organized wage disbursement of both permanent & temporary workers coupled with procurement of the commodities from the estate control shop. Also it was possible to verify the data by cross verification and witnessing the transactions. Direct interview formed one major source of collection and verification of information. On the contrary, at Shenbagathope systematic working pattern by the Paliyans was nearly absent. They did not receive their sale proceedings in a regular channel. Again return from sale proceedings depended upon various factors such as efficiency of the seller, disposal of the products to appropriate buyer, genuine prices etc. Despite these problems an effort was made to obtain accurate informations. The well off family in Palamalai earns through cultivation of coffee, ginger etc which amounted to Rs. 10477/ - in the year 1992 – 93. This family also earned Rs. 3840/- through the collection & sale of Minor Forest Produce. In Shenbagathope no Paliyan owned land. While the field work was going on it was found that Mr. Mutheyas family (well – off family) worked arduously and spent their money with lot of planning to save as much as possible directly or indirectly.

In both the village rice forms the staple food and therefore it is the major item of expenses followed by other items. Jaggery or Palmyara sugar is the second most necessary item. Coffee is collected from individual farm or from the estate and is taken twice a day. Though alcohol is consumed by members of both communities there is a marked distinction between the two. In Palamalai, on Saturdays which is the wage day of the week, alcohol is available on the outskirt of the estate. However it is available on other days also, but few consume it. Seldom, any regular worker stops the work due to drunkenness. While in Shenbagathope if one has sufficient cash he spends the amount and go to work only when it is finished or on the verge of being finished. Palani who was appointed as fire watcher spends more than 60% of his salary on alcoholic drinks every month. But in Palamalai there is social control due to discipline enforced by missionary fathers. In Palamalai saving habits, exist in different ways. One is private chit fund. Selvaraj & his wife narrates that it is only through chit fund that they have been able to buy their major household articles as they get the amount in lump sum. They usually receive one saree / Dhoti from Govt. annually free of cost. They also purchased one to two sets for themselves. Apart from that they supplement their needs by purchasing old cloths from the traders. Spending a large amount on transportation showed their inclination to travel and enjoy. In Palamalai people walk down 10 k.m. to catch busses. They go to hospital, pilgrimage centre (Viz., Velankani temple, etc.) nearby auctioning centre and to their affinal relation houses etc. in bus. Vyakulamary (of the 2nd family) in the Palamalai village is married near Palani. Both visit their village in a fortnight. About the use of condiments, it is seen that at Palamalai Paliyans prefer to take each items to make different delicious food. On the contrary, at Shenbagathope people simply purchase powdered spices (i.e., mixed powdered condiments). It may be observed that at Shenbagathope the transition to prepare different food items is of recent past. It is evident from the data collected that in both the villages the dependency on forest products, for food, has been reduced substantially. Though at Shenbagathope dependence on forest products for subsistence was higher earlier it has decreased due to increase in income. Not far away from Shenbagathope is Pilavakkal Dam settlement. Paliyans at Pillavakal Dam settlement earn less compared to that of Shenbagathope. Usually their dependence on forest and other foods are equally divided. They take at least one or two meals a day and live on yams and tubers on other times.

**Table No. 3.10 – Weekly activities two household in Pilavakkal Dam settlement**

| Velandi | | Mutheya | |
|---|---|---|---|
| 1st day | Collection of Tuber | 1st day | Rs. 39/- from fire wood |
| 2nd day | Went out | 2nd day | involved in collection of *Indum Pattai* & fire wood |
| 3rd day | Did not do any work | 3rd day | Selling of the bark of *Indum Pattai* and fire wood |
| 4th day | Sold *Indum Pattai* for Rs. 20/- | 4th day | Collected *Indum Pattai* and removed the bark |
| 5th day | 10kg. 1. *Indum Pattai* sold for Rs. 40/- From five wood received Rs. 13/- | 5th day | Obtained two rabbits in the traps. |
| 6th day | Nill | 6th day | Collection of fire wood. |
| 7th day | Weeding agricultural field for Rs. 30/- per day | 7th day | Collection of fire wood. |

They set out to forest with left over foods from the previous night, collect bark and wood etc. (fire wood) especially *indampattai* for sale nearby. While in forest they collect tubers and yams and consume it.

## Discussions

In Chapter – I, an effort has been made to explore and identify the nature of traditional culture of the Paliyans including ecology, technology and subsistence pattern based on hunting and food gathering activities. In Chapter – II we have discussed how there had been changes in ecology and introduction of coffee plantations, orange and guava orchards, commercial crops such as black pepper, cardamom, ginger and other exotic vegetables are grown. Such innovations have tremendous impact on traditional economy which modified the economic pursuits of the Paliyans and affected significantly their style of life and behaviour. Yet different localities of the Paliyan habitat differ from region to region. Here we have tried to show the economic changes in two villages in order to visualize the process of change in the Paliyan habitat in different localities and income, expenditure and changing subsistence pattern.

With a view to show changes in Malpandaram economy Brain Morris has developed a model of four stages on the basis of different categories of villages. (Morris: 1982:62). The author had tried to modify that model to suit to the Paliyan situation with some elaboration.

Stage I: The traditional pattern of hunting and food gathering, a completely nomadic life style. The exchange of forest produce was mainly local and conducted mostly through barter.

Stage II: Roughly equivalent, but at this stage the individuals have links with some settlement. The forest contractors appeared in the scene and exchange of MFP was channeled through the forest contractors.

Stage III: Here the dual subsistence pattern breaks down and agricultural labour supplements the collection of food resources. The collection of minor forest produce is the primary economic activity, residence patterns are largely sedentary though individual families may occasionally change their residence within a cluster of settlements where extended kin groups reside.

In Pilavakkal Dam area though forest is near there are very limited tamarind trees, mango trees etc. Most of the trees are gallnut, 'Indampattai' etc. Paliyans here are specialized in honey collection. Throughout the year they supplement their food with different kinds of yams. Other than minor forest produce collection, selling of fire wood is another major source of income. Usually they sell the 'Indiampattai' bark and the wood portion also.

**Table No. 3.11 Collection of M.F.P. in two traditional settlements at Shenbagathope & Pilavakkal Dam**

| Sl.No. | Month | Name of the M.F.P. | |
|---|---|---|---|
| | | Pilavakkal Dam | Shenbagathope |
| 1. | '*Tai*' December – January | '*Indum pattai*'*(Acacia Instia)* 'Galnut' Honey | Tamarind |
| 2. | ' *Masi*' January – February | '*Indum pattai*' *(Acacia Intsia)* 'Galnut' Honey Tamarind | Tamarind Mango |
| 3. | 'Pangkui' February – March | '*Nikunta Kai*' *(Marking nut)* 'Indum pattai' *(Acacia Intsia)* Honey | Tamarind Mango |
| 4. | '*Chitrai*' March – April | '*Nekunta Kai*' *(Marking nut)* '*Indum pattai*' *(Acacia Intsia)* Honey | Tamarind Mango |
| 5 | '*Vaigasi*' April – May | Full wood | Tamarind Mango Guava |
| 6 | '*Ani*' May – June | Fuel wood | Tamarind Mango Guava |
| 7 | '*Adi*' June – July | Full wood | Tamarind Mango Guava |
| 8 | '*Avani*' July – August | Fuel wood | '*Kungulium*' *(a resin)* '*Sadikapoo*' |
| 9 | '*Punatachi*' August – September | '*Kungaliyom*' *(a resin)* ' *Nellikai*' *(Goose berry)* | '*Kungaliyom*' *(a resin)* '*Sadikapoo*' |
| 10 | '*Aipasi*' September – October | '*Lavanga Pattai*' Wood | Agricultural labour |
| 11 | 'Kartikai' October – November | '*Popodatum (Rs. 15)*' (Herbal medicine) 'Sundakai' | Agricultural labour |
| 12 | 'Margali' November – December | | |

**Stage IV:** An advanced stage of adaptation, the collection of forest products is nebgibible and in their subsistence pattern the Paliyans are equivalent to neighbouring peasant cultivators.

The income (approximately) and expenditure of two sample families of Palamalai & Shenbagathope are presented above. In

Palamalai, besides being agricultural labour in the missionary estate, some people have their own agricultural plots and garden as mentioned earlier. Paliyans in this village can be placed in between stage three and stage four category.

The Paliyans at Shenbagathope appears to be in a transitional state form stage – III to stage – IV. Their major economic activity are mostly confined to collection of MFP. They do not possess any agricultural land nor they have employment in organised sectors as a major means excepting a few (engaged recently in flying squirrely sanctuary).

The ideal pattern of stage – III is reflected in Pilavakkal dam Paliyan settlement, a neighbouring hamlet of Shenbagathope. Gardener's observation regarding Paliyans that they are oscillating between a traditional state and culture contact situation is changed at the present study. It has been found that Paliyans at Palani hill are wavering between a culture contact

CHART NO. - 3

ECONOMIC LIFE OF PALIYANS IN VARIOUS STAGES

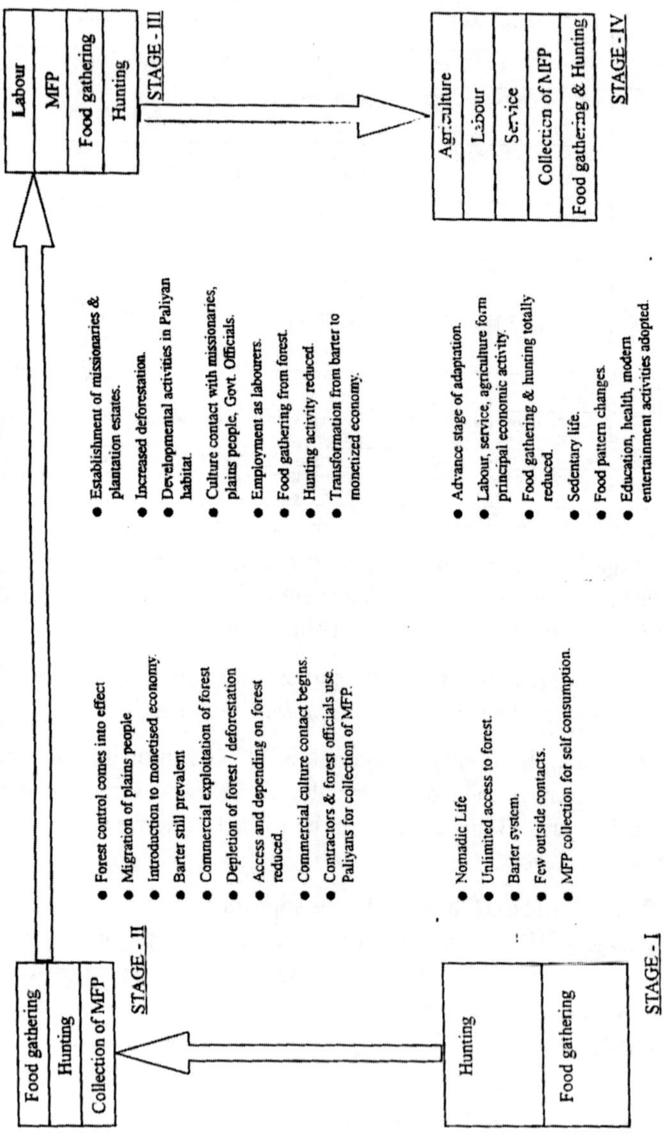

situation and modern life. Even though they have not yet accustomed to the latter, they have left their traditional or primitive economic life. Their staple food is no more yams or diascorea. Close observation indicate they have changed their food habits use dress and household appliances as in other parts of plains area. They have left nomadic life, and adopted sedentary life. Their principal sources of income comprises labour in organised or unorganised sector collection of MFP and few in regular service sector. Foraging and hunting has been pushed to a state of leisure activity. They have adopted modern entertainment sources like cinema, radio rather than confined to own community dancing and singing, travelling etc. Comparatively Paliyans at Palamalai are in more advanced stage of adoption than Shenbagathope or Pilavakkal Dam settlement, owing to the influence of missionary fathers. Paliyans at Pilavakkal Dam settlement are more traditional, but also lead a sedantory life and still depend upon forest for their living. Lack of outside economic opportunities nearly slowed their transformation process and preserved their isolation to some extent. But they are not averse to adoption.

Socio-political consciousness has been on increase and gradually the feeling of assertion is visible in the attitude and manners. Forest and economic life of Paliyan were intertwined at Shenbagathope and Palamalai.

Restrictions on forests and increased deforestation has snatched their forest dependency living pattern and therefore Gardener's observation looses its relevance today. Present day Paliyans have more or less adopted to monetized economy and their transition and transformation can be best understood by Morri's model described above.

## Notes

1. In Pallamalai at one side there are a no. of megalithic monuments which indicate its long distant antiquity.

# 4

# Changing Ecology and Economy

## SECTION – I

## FOREST POLICIES DURING COLONIAL, PRE-COLONIAL AND POSTCOLONIAL ERA.

" Two thousand years ago, as much as 85% of the Indian sub- continent was covered with forest. By 1990 less than 10% of the country possessed good forest cover" (Poffenberger M. et al; 1996:2). The above statements are sufficient to substantiate the changes that has occurred in the ecology, leading to the changes in economy. "Though both the terms ecology and economy are rooted in the same Greek work 'oikos' or 'household' , yet in the context of market oriented development they have been rendered contradictory". (Shiva V; 1991:30) Ecological destruction is an obvious cost for economic development. In the present chapter it is difficult to enumerate changes that occurred during the last 2000 years, but never the less effort has been made to present the changes that has occurred during colonial rule in this sub-continent based on the available literature followed by observations in regard to Paliyan scenario.

Everybody is dependent on natural resources in day to day life. With advancement of science and technology man has made efforts to have a control over it. So there was some kind of control over vegetation, soil and water at different stages of history. But we will concentrate on the colonial period as the base line for such debate & also see the imperial attitude towards the natural world and strategies of underprivileged to the new system of resource extraction & regulation. Though colonial era is considered to be the benchmark towards penetration of market forces by many analysts leading to commodification of natural resources, there are instances to disprove this. All such discussion would take place in brief and some times not in a complete way, as the basic objective is to see the trend of impact on resources.

Water which is fundamental to the survival of living being is also under strain. Our cultural heritage perceived water as the basis of all life having a complex ralationship with soils, plants and human needs. Large scale water projects which work against natures water economy and people's sustenance requirement have been designed by destroying water technologies which were ecologically more sustainable and socially more just.

The irrigation works in Pre-British India were managed by a variety of social organisations within the village community. When the Britishers came to India they had no expertise in water management, since agriculture in Europe has never been dependent on irrigation.

The Britishers, however, had to control over river water in India keeping in view their economic interest in increasing revenue. In Rajasthan they controlled water to maximise the revenues derived from their salt trade, in Bengal to protect their transport networks & increase revenue from agriculture (Shiva V. 1991: 191). Similar trends of control mechanism was adopted during colonial period which had a direct or indirect impact on forest resources. Though Anthropologists have documented the disruptive impact of modern forestry showing the harmony between the tribal communities and the environment, this view ignores the effect of changes over time and offers an overly functional explanation of human behaviour. And sometimes these studies have lacked, when and why there was a shift in official policies and attitudes.

Their underlying assumption is that traditional system of land use were inherently stable and the disruption was on account of external pressures.

In the Pre-British era the Mughals perceived a civilized society as one primarily engaged in agriculture. So there was an assumption that the prosperity of the Mughal ruling class hinged upon its success in expanding and strengthening the empire's agrarian foundations. States such as the Delhi Sultanate under Mohammed bin Tughlaq (1325-51) tried to encourage agrarian extension by remitting revenues & by providing credit to those peasants who cleared fresh land for agriculture, Woodcutter and Ploughmen accompanied the Mughal army from the time of Shah Jahan (1628-56) and there was a general order that who so ever cleared a forest (brought land under cultivation) such land should be Zamindari. Abul Fazal (1873), writes "the 'amalguzar' should strive to bring waste lands into cultivation and take heed that what is in cultivation fall not waste". Forest lands were destroyed to win over wars. In Bihar Shah Bajkhan who was assigned the task of suppressing the rebellion of Gajapathi a zamindar of Jagdishpur in Bihar had to spend nearly two months in cutting down the trees around the fort before he could capture it (Chenan Singh: 1995). In the pre-colonial indigenous economic process, the levels of utilization of natural resources were not significant enough to result in drastic environmental problems (Shiva: 1991:16). But it was only during the British colonialism that a systematic approach was adopted to utilize / exploit the natural resources obtained from the forest.

The debate over the distinction between pre colonial – colonial period is worth mentioning at this juncture. Guha and Gadgil argue that in the earlier situation village communities had control over management and disposal of forests and uncultivated lands. Demands by dominant land holders and rulers were limited and never approached the scale they did in the subsequent period of colonial rule. On the contrary, Grove states that deforestation had assumed significant proportion before the advent of colonialism. He also argues that initiatives for state control of forests in Sindh anticipated colonial conservation (Mahesh Rangarajan. 1996 :16). Thus stock keepers and animal trappers, shifting cultivators and itinerant communities were never subject to the kind of control that was to be attempted by the British. A functional explanation would be misleading, but the British had a more antagonistic attitude to mobile groups.

The new ruler's concept of political power was more universalist than that of indigenous rulers. So it is necessary to understand why Britishers were acting faster in controlling forest resources. The strategic and commercial imperatives of a transcontinental empire demanded a large fleet resulting in increasing demand of wood. In Bombay & Calcutta ships were built & slowly, this led to shortage of wood. The decision to have control over teak led the first step towards control of forest. A conservator of forest was appointed in Malabar in 1806, but this was resented by locals. Guha (Guha: 1990) states that the term annexationist has been used to describe the views of officials who advocated total state control of wood lands during the debate on forest acts between 1865 and 1878. But an opposite stand was taken by Thomas Munro who advised on free market & minimal role of Govt. Citing how the residents in Malabar planted and protected trees on a regular basis, he bestowed his complete faith in the ability of peasants and other land owners, to act in their own long term interest. This led to abolition, of the post of conservator in Malabar in 1823. Stebbing writes "the abolition of the post was more beneficial to traders as there was unfettered access to woodlands.' But then the debate on conservation was revived by the surgeons Alexander Gibson and Heigh Clegharn. They linked deforestation to drought. Kumri shifting cultivaion in the Wesern Ghat was focussed as a vital case for substantiating deforestation. But then the ulterior motive was to persuade the hill people, or groups living in the fringes of cultivated, arable, zones to take up plough cultivation or wage labours, (Bayy, Indian society P.141) as there was a need to augment revenue & labour supply. The pressure on the forest increased as there was need of wood for British armoury. New policies were thought for renewal. Lands were acquired on lease basis & restriction on use of forests for promoting growth of timber were implemented. The Madras Government acquired the Annamalai forests in 1847 & sought to protect teak from fires (Brandis, D: 1897). In 1857 the Resident at Lucknow Lt. Col. William Sleeman ordered that the best forests be acquired on lease and trees reserved for ordinance factories (NAI, 1944). So Indian forest Department was set up in 1864 to have a control over the forest resources. This lead to transformation of common vital resources into commodities for generating profits and growth or revenue.

With the establishment of British colonial rule in India the ever-increasing revenue demands of the industrial revolution in England were largely met from this country. Forced cultivation of Indigo in Bengal and Bihar, cultivation of cotton in Gujarat and the Deccan etc. led to large scale commitment of land for the supply of raw materials for the British textile industry, the flag bearer of industrial revolution. Forests in the sensitive mountain ecosystem like the Western Ghats or the Himalayas were felled to build battle ships or to meet the requirements of expanding railway networks". (Shiva 1991 :16).

A brief review of the forest policy would be mentioned here to show the changing trend over time.

## Forest Act 1865

When the first forest Act was passed in 1865, the entire India had come under British sovereignty. Under this act Government could declare any land covered with trees as government forest & formulated rules for their management. This Act came into effect on 1st May 1865.

## Indian Forest Act 1878

This new Act replaced the forest Act 1865. Here government had classified the forest into three types Viz:-

1. **Reserve Forest:** These forest areas are permanently retained as Government forest and the local people are not given any rights over this area.

2. **Protected Forest:** In these forests neighbouring people have certain rights, but they cannot deforest it without the prior permission of Forest Department who are also authorized to close down any such forest for a period of maximum 30 years.
3. **Village Forest:** For meeting the needs of villagers, Government declared certain areas as village forest where the villagers have right over it. Government is empowered to cancel reserve forest to make it village forest.

This process had its impact on various section of the society as is evident in the writing of social thinkers like. Jyotiba Phule (1882).

In the past the peasants who had small pieces of land who could not eke out enough from it for their survival used to eat fruits from the nearby forest and used to collect leaves, flowers and dried tree branches & by selling these to others supplemented their income. They also used to maintain a couple of cows or goats and were living happily in their villages depending on the village common grazing land. But her majesties Govt's conspirational bureaucracy have used their foreign intelligence and have newly established the great forest department & have incorporated all mountains, hills, valleys, alongwith barren lands and village common grazing lands, in this department, thus making it impossible for the goats of the poor peasants to find even breathing space in the forest *(Hiremath S.R.: 1995: 2).*

## National forest policy 1884

This policy emphasized on exploiting the forest resources for increasing the state revenue. In this policy government brought out the rules and regulation about the rights and privileges of forest dwellers over forest land & its produce.

The forest policy of 1894 placed the needs of agriculture and animal husbandry as the first priority followed by local requirement and needs. Forest departments and officials appeared in the scene. Tribal people who hitherto enjoyed unlimited rights inside the forest faced restrictions and regulations. This hurt and affected them deeply. The 1894 policy lacked equilibrium between various groups of user's of forest resources. (Mehta B.H., 1968:211 & Elwin V, 1963:52).

## Indian Forest Act 1927

It contained almost all provisions of the earlier Acts except for some changes like duty on timber as Hiremath narrates.

"It is stated in very beginning that (it is), an Act to consolidate the law relating to forest, the transit of forest produce and the duty leviable on timber & other forest product". (Hiremath S.R. et al: 1995:18).

With the introduction of 1927 forest Act, village & tribal communities were debarred from their traditional rights in the forest, resulting in anger and frustration. Simultaneously another development was taking place during colonial period that had an adverse effect on the forest. Geological surveys led to the exploitation of coal & minerals in vast forest areas. Botanical survey revealed the economic utility of tree wealth. (Mehta. 1968: 210). The exploitation of the former needed construction of roads and railway lines for transportation, resulting in clearing of valuable forest. Again this also resulted in inflow of nontribals to tribal habitat. In many places forest areas were cleaned up for construction of industries.

The economic importance of plant kingdom in the forest environment resulted in commercial forestry. So extensive plantation of monoculture species look place.

"The imperative for increasing revenue in a growth economy, however drives resource utilization pattern in a direction which maximizes production of commercially valuable components of the ecosystem at the cost of destruction of those components which are commercially value less but essential to survival". (Shiva V. 1991: 80).

As mentioned earlier, in many areas, portions of natural forest were given on lease basis. So in the name of natural forest tea, coffee and cardamom plantation came up. Again leased natural forest products were given at a throw away prices in the name of promoting industries.

## Forest and tribal's in the post Independence period

It has been observed, how in the colonial era, a strong force brought about radical changes in the interaction of tribals with their forest. In the year 1952, the first forest Act of independent India was enunciated. In this act six important needs of the nation was highlighted viz.

(i) The need for evolving a system of balanced & complementary land use

(ii) The need for checking.

  (a) Denudation in mountainous areas

  (b) Erosion in the treeless banks of great rivers.

(c) Invasion of sea sands in the coastal tracts and shifting sand dunes in the desert.

(iii) The need for establishing tree lands for ameliorating physical and climate condition.

(iv) The need for ensuring supplies for grazing, small wood for agricultural implements etc.

(v) The need for sustained supply of timber and other forest produce required for defence, communication & Industry.

(vi) The need for realization of the maximum annual revenue. (Roy Burman B.K. 1968:170)

It is obvious that our leaders during the era of struggle for independence could visualize the needs of their tribal brothers in a more rational & practical way. Unfortunately, this did not happen during post independence period. The lively example is the 1952 Forest Act. In the name of scientific forestry & to raise the forest cover from 20% to 33% certain basic needs of the tribals were ignored. This new policy of independence era even deviated from the earlier forest Act in protecting the interest of forest dwellers viz.

"(i) The old policy envisaged the release of forest land for cultivation subject to certain safe guards.

The new policy withdrew this concession.

(ii) The old policy had left margin for the supply of the villager's need from the outlying areas in the reserved forests. The policy decided that there should be village forests for the purpose.

(iii) The old policy did not touch the private forest of the tribals. The new policy applied the same control to them.

(iv) The old policy did not touch free grazing in the forest, the new policy sought to bring it under control. Fees were introduced and grazing was to be kept to the minimum. (Roy Burman 1968: 204).

The policy of increasing the forest cover from 20% to 33% resulted in various problems. In M.P., Bihar & Orissa some of the unsurveyed land (including agricultural land) which were earlier 'zagirdari' land were converted to reserved forest. Maharastra forest Department claimed areas without any plants as reserve forest. In West Bengal tribals had certain rights in exzamindari land, which was curtailed. These indiscriminate policy of forest planners led the 'Scheduled Caste & Scheduled Tribe commission' to write.

In the circumstances we are obliged to suggest a new approach. The forest department as a branch of government should consider itself responsible for the development of tribals as well as the forest areas. It should not think of an exclusive obligation to trees or vegetation, but should also accept an obligation to utilize the forest as much as possible from the angle of economic development of the tibals residing in it. In return, the tribals should recognize the need to reserve the forests. An essential feature of this new feature is that the forest Department should under take the responsibility of giving adequate employment to the tribals living in and around the forest areas development of the forest and the betterment of the tribals should no longer be at logger heads. They have to go hand in hand. A close study of the problem from this point of view will open out before them a great vista of hope in place of the present despair and the tribals will find before them an endless panorama of employment opportunities. This calls for a partnership rather than an exclusive approach which arises from the policy enunciated in 1894 & 1952 and the manner in which it has been implemented. The first step in the direction should be a co-ordinated approach to provide employment opportunities in forest, agriculture, animal husbandry and industries for at least 300 days in a year. (Hiremath 1995: 18).

There are some analysts who feel that the 1952 forest policy was narrowly interpreted. The construction of roads, irrigation & hydro electricity projects were some time taken up ignoring the ecological sensitivity of an area and these were justified in the name of national interest, while ignoring the needs of forest dwellers.

Several oppositions to this attitude based on factual problem were brought to the limelight. These included Government reports, academic articles, N.G.O. movement and tribal movement etc.

## Indian Forest Bill 1980

On the recommendation of National Commission on Agriculture a Draft forest Bill was prepared by the Department of the forest in the year 1980. This bill was considered antipeople and it was felt that it discouraged the use of forest land and its produce by the poor. To add to the problem more, the forest officials were granted special powers. "Forest officers were given very wide power of arrest and seizure of property to deal with the offences suspected to be committed in respect of forest" (Hiremath S.R. et al 1995). But due to strong opposition this bill was withdrawn.

## Forest conservation Act 1980

This was the first Act after the subject of the forest was transferred to the concurrent list of constitution in 1976. The salient feature of this Act is that state government is prohibited to declare any reserve forest or any portion thereof, non-reserved without prior permission of the central government.

## National Fprest Policy 1988

Govt. of India adopted in 1988 the forest Policy Resolution which recognized the rights of the people over forests. "It was clearly accepted that all efforts to conserve and develop forests must be based on people's participation" (Bhat C.P. 1995: VII).

This new Act of the Government of India was presented in Parliament in December 1988, replacing the earlier Act of 1952. Government admitted that the basic objective of 1952 forest policy, aimed at forest coverage of one third of the total land area, had failed miserably along with destruction of flora and fauna. So the minister while presenting the new policy out lined some of the salient feature of the same i.e. the new forest policy of 1988 as follows.

1. Maintenance of environmental stability through preservation and restoration of ecological balance.
2. Conservation of the natural heritage of the country by preserving, the remaining natural forest and protecting the vast genetic resources for the benefit of the posterity.
3. Meeting the basic needs of the people especially fuel wood, fodder and small timber for the rural and the tribal people.
4. Maintaining the intrinsic relationship between forest and tribals and other poor people living in and around forests by protecting their customary rights and concession on the forest."(Hirmath S.R. et al 1995: 144).

## Joint Forest Management (J.F.M)

With the noble idea of involving tribal and other rural people living in and around forests in protection, regeneration & development of forest, in the New forest policy of 1988, a new chapter on India's forest environment has been opened. On the basis of this ideology, on 1st June 1990 Government of India issued a circular facilitating the involvement of N.G.O. and local communities in the development of degraded lands while meeting the needs of the later. This epitomizes a sudden departure from the earlier stand of victimizing the local communities as the culprit for the destruction of forests.

This seems to be well accepted going by the statement of Hiremath S.R. (1995) that "Now more than 15 states have passed government orders formalizing joint forest management."

## SECTION-II

## CHANGING ECOLOGY-GROWTH OF PLANTATION AND URBANIZATION IN PALANI HILLS

In the previous section changing ecology at national level and the present debate on forest policy of British administration has been discussed. In the present section, the ecological changes in and around the areas inhabited by the Paliyans have been presented. The succeeding section discusses economic changes of the Paliyans occurring due to such ecological transformations in the region. Ecological changes around Paliyan habitat are observed to have occurred due to three reasons.

(a) Urbanization in Palani hills

(b) Growth of Population due to migration

(c) Growth of plantation.

Palani hills which accommodate major concentration of Paliyans provide excellent scope concerning this study leading us to important and significant conclusions. Srivilliputhur is the other inhabitation studied.

**(a) Urbanization in Palani hills:** Commercial exploitation of Palani hills commenced during the colonial period U.B.S. Ward mentions that till that time the virginity of the hill remained undisturbed. He was the first foreigner to set foot at Kodaikanal to survey Palani hills (Wyck C.C:1945).But credit for emergence of a settlement in Kodaikanal leading to a scenic hill station goes to the American missionaries. In those days they used to go to Jaffna (Srilanka) which was utilized as the summer resort in order to escape the heat of Madurai (South).In the year1838, the mission made a proposal to acquire a vessel to carry the sick back and forth from Jaffna. The proposal was rejected with a suggestion to construct a sanitarium at Sirumalai hills (an offshoot of Palani hills). But the persons engaged suffered from malaria during construction and the project was abandoned. The search was on to relocate the sanitarium at a better place. At this state Rev. C.F. Muzzy and others explored the top of Palani range and found it to be an ideal place for sanitarium. Thus the Missionary centre was developed in Kodaikanal. In the year 1846, a French coffee planter, Mr. M.F.

de Fonclair began planting coffee in lower Palni hills and became the first outsider to take up permanent residence in Kodaikanal. This was followed by growth of Australian blue gum and other trees in the Bombay shola and laid out what was town's first garden by Major J.M. Patridge.Government was approached to allot land for construction of houses.At this stage the Government decided to extend the Nilgiri hill range Act applicable to Palani hills. As per this Act, land was available on rent / lease at the rate of five rupees for the first *kani* (equivalent to 1. 32 acres) and two rupees eight annas for each additional *kani*.

The American mission brought Bombay Shola from Major Patridge for rupees one hundred fifty four (Bruce De Zong: 1971). Similarly the missionary estate at Palamalai was bought for Rs. 8000/- (Dairy of General Procurator: 1904). In 1863, the Collector of Madurai, under whose jurisdiction the area belonged, intended to develop Kodai lake to add beauty to the hill station. Col Law came to Kodakanal in 1875 and he completed in three years what is known as Laws Ghat Road. With communication opening up, the small town grew in size and Kodaikanal Municipality was formed in 1899. Work on forty mile long round road began in 1929. This was completed in three years in 1932. Subsequently another road to Kodaikanal known as Munnar road was opened in 1942 and road link from Kerala was made available. These multiple road links provided an easy access to the hill area and developed with an intention of using them as alternative escape routes for the British army (during World War II – operation).

In the post independence period, Government implemented various developmental schemes. Priorities were given to construction of roads in the region specifically under "Western Ghat Scheme". In 1976, road from Perumal Malai to Palani was opened. Thus access was available to Palani hills from all sides, which helped, increase in communication and made the contact situation more frequent. Kodaikanal attracted tourists from beginning. With better roads people criss crossing caste and linguistic barriers started flowing into the hill station. The physical terrain that was working as a barrier between Paliyans and the outsiders no longer existed. In course of time many educational institutes, hotels, holiday homes grew up. In addition, Palani Hills attracted Christian communities because of presence of missionary. In recent times, migration of labour force, rehabilitation of Srilankan refugees etc. contributed to the existing population and resulted in reduction of physical and social isolation of Paliyans. In Palamalai estate, we came across the missionary management where people from different communities assembled together. Missionary fathers sometimes encouraged inter-community marriages, which includes Paliyans. This trend is not limited to Palamalai alone. It extends to the neighbouring areas where missionaries are operating. Improvement of communication contributed to accelerated urbanization of the Palani Hills.

**(b) Growth of population due to migration:** Gazetters indicate that the process of migration took place some five hundred years before (i.e. much earlier to the colonial rule) Documents of the colonial period mention existence of Puliyans as well as Paliyans as the earliest inhabitants of Palani hills. The latter worked as serfs in the fields of migrants while the Paliyans remained more or less independent.

During British rule, the East India Company introduced "New Land Revenue System". Prior to British rule the village ownership of land existed, the village was the unit of assessment and the payers of revenue. But the East India Company abolished the earlier revenue system in which the village was the unit of assessment and payments. Instead they created individual holders of land and introduced the system of individual land assessment and revenue payment. (Desai: 1987: 43).

Further, for smooth collection of land revenue Zamindari system was introduced. Zamindars not only collected land revenue but also exploited the poor peasants ruthlessly. The land revenue was raised gradually tenfold and due to oppression of Zamindars, peasants in rural India left their hearths and homes and migrated to other areas in search of peaceful life. In Tamilnadu also rural people under the pressure of new land revenue system tried to migrate to the Palani hills where British Govt. had decided to extend the Nilgiri Hill range Act applicable to Palani Hills (Ref. urbanization process mentioned earlier in this chapter). Due to easy availability of land on lease and sale, middle class entrepreneurs of the plains area were attracted and migrated to Palani hills and tried to grab the forest that provided subsistence for the Paliyans. The 'Nadars' (toddy tapers) of Tirnelveli, 'Naikars' and other militant caste groups of Andhra migrated to Tamilnadu during the rule of the 'Nayak dynasty in Madura. People from

Karnataka and the Muslims from Madurai also populated the Palani hills, which indirectly affected the economic well being of the Paliyans.

Besides, British Govt. Introduced monetized market economy and implemented new Excise Acts by which the dealers of excise goods became authorized dealers of commodities that was strictly controlled by the government. These new entrepreneurs substantially changed the economic structure of the Palini Hills and provided new opportunities for the adoption of new economic system. Such situations were thoroughly exploited by the migrant rural population who could manipulate the new economic situation while the Paliyans remained deprived from the newly available opportunities in the changed circumstances.

Though during the colonial period there was no scientific study of urbanization and growth of population concomitant with the ecological hazards, the recent study by Sukumar et al highlighted such situation through the scientific study of such changes during recent period (i.e. from 1961 – 1981).

Sukumar's unpublished paper state "Kodaikanal taluk area has increased nearly four times from 1961 to 1981. This has taken place at the expense of the forest in the Palani Hills." Further the table below (as given by Sukumar et al unpublished paper) provides an insight into the population growth pattern of the area.

**Table 4.1 – Decennial variation in population growth, houses, households and total workers.**

| | 1961 | 1971 | % variation | 1981 | % variation |
|---|---|---|---|---|---|
| Total Population | 50710 | 71453 | 40.9 | 83585 | 16.9 |
| Residential Houses | 10882 | 15274 | 40.3 | 15333 | 0.4 |
| Houses Holds | 11247 | 15719 | 39.7 | 18548 | 17.9 |
| Total Workers | 24229 | 29302 | 21 | 38222 | 30.4 |

Missionary activity, Plantations, tourism and other economic and administrative activities contributed to the growth of population and increase of size of the Kodaikanal town. In order to augment tourism, hotels and private resorts were built. Transportation and other commercial activities increased; all at the expense of forest bit by bit. This however, reduced social and physical isolation of the local inhabitants particularly Paliyans. Along with ecological changes, the economic activity around the hill also has undergone transition during these periods.

## Growth of plantations in the Palani hills

The above brief account is a bird's eye view of the radical ecological changes that has occurred in the Palani hills over the years. But then, what is also significant to note is that this virgin forest region was more systematically exploited during the colonial era. The introduction of coffee plantation in the mid 19th century necessitated the cleaning up of forest land. Thousands of acres of fertile land were converted into plantations to grow cash crops like coffee, orange, jackfruit, cardamom, black pepper, ginger etc. Prior to the introduction of these new species farmers were cultivating different kinds of millets. But realizing the economic potentialities of cash crops most farmers abandoned the cultivation of millets etc, which were not economically profitable. So there was change in the cropping pattern. Further new plant species like Eucalyptus, Australian Blue gum, etc were grown.

The introduction of plantation of wattle, eucalyptus, pine, silver oak and other horticultural crops reduced the natural shola forest and native grasslands which were the abode of the numerous wild life. Plantation owners, being wary of wild animals and some times as a leisure sport, hunted the animals reducing their population. Reduction of forest cover and availability of horticultural products in the plantation areas allured the wild animals. Plantation owners are provided with fire arms either for intimidating wild animals or use in self defence. But these arms are more often utilized for hunting these animals than the specified purpose. Many endangered species have been lost with depletion of forest cover forever. Natural springs have been lost and created water scarcity in the summer due to receding water level.

Construction of road ways led to large scale cutting down of trees, which further led to land slide and soil erosion at many places. During rain, the top soil eroded without any support and led to land deterioration making certain areas barren.

Sukumaran et al (unpublished paper) writes "with the interventions by the people most of the springs have become extinct. Acidity of the soil has increased by affecting the productivity of the hill produce."

Under these changing circumstances Paliyans are left with limited choice. Dependency on the forest product has turned out to be an uncertainty. Forest laws are not in their favour which made them vulnerable to the loss of their traditional mode of living.

## Scientific study of changing ecology

The scientific studies undertaken so far corroborate the foregoing observations.

A study was conducted by Sukumar, et al, (1985) using multidate LAND SAT FCC imagery (154-053 on 1:2,50,000 scale) of two periods February 1973 and January 1982 and mapped land use/land cover categories like semi evergreen forest, mixed forest, dry deciduous forest, scrubs, horticultural crop lands, mixed plantations, orchards, barren grass lands & forest lands through visual interpretation techniques. The study concludes that there has been environmental disturbance which have occurred in these areas, and have been broadly classified into four categories on level – I and seven categories in level – II (as shown in the following table).

**Table No. 4.2 – Land use / Land cover classification for Palani hills based on Visual interpretation of Land Sat imagery**

| Level I | Level II | 1973 | 1982 |
|---|---|---|---|
| 1. Forest | a) Semi evergreen<br>b) Mixed forest<br>c) Dry deciduous<br>d) Scrubs | 60.89% | 55.82% |
| 2. Agricultural Lands | a) Horticultural crop lands<br>b) Mixed plantation<br>c) Dry deciduous<br>d) Scrubs | 26.96% | 32.42% |
| 3. Barren / Grass lands Forest blanks | | 12.08% | 11.62% |
| 4. Water Bodies | | 0.07% | 0.13% |

The table no. 4.2 and table no. 4.3 indicate that the forest cover have been reduced from 93586.11 hectares (60.89%) to 85759.75 (55.8%) or by 7826.39 hectares (5.09%) during a span of 9 years between 1973 and 1982. In contrast, horticultural and mixed plantation has been on increase from 36579.19 hectares (23.80%) to 44,982.24 (29.27%) an increase of 8403.05 hectares or 5.47% during these periods. This can lead us to imagine the extent of deforestation and growth of plantation till date. The impact of this ecological change over the years on Paliyans has been a slow and gradual process.

**Table 4.3 – Area and Pattern of Change of Land Use/Land cover for Palani Hills**

| SL. No | Land Use/ Land cover Level – I | Level – II | 1973 | | 1982 | | Percent variation |
|---|---|---|---|---|---|---|---|
| | | | Area in hectare | Percent | Area in hectare | Percent | |
| 1 | Forest | a) Semi evergreen | 4.995,98 | 3,25 | 4.622,86 | 3,04 | -0,21 |
| | | b) Mixed forest | 40,202,72 | 26,16 | 35.923,20 | 23,37 | -2,79 |
| | | c)Dry deciduous | 24,009,28 | 15,62 | 22.247,49 | 14,47 | -1,15 |
| | | d)Scrubs | 24.378,13 | 15,86 | 22.966,20 | 14,94 | -0,92 |
| 2 | Agricultural land | a)Horticultural crop land | 11.539,13 | 7,51 | 16.493,57 | 10,73 | +3.22 |
| | | b)Mixed plantation (Mainly coffee) | 25.039,86 | 16,29 | 28.488,67 | 18,54 | +2.25 |
| | | c)Orchards | 4.854,03 | 3,16 | 4.854,03 | 3,16 | 0,00 |
| 3 | Barren/ Grass land/ Forest Blanks | | 18.567,48 | 12,08 | 17.855,27 | 11,62 | -0,46 |
| 4 | Water Bodies | | 112,95 | 0,07 | 206,47 | 0,13 | +0.06 |
| | **TOTAL** | | **153.699,76** | **100,00** | **153.699,76** | **100** | |

**Source:-** Area calculated from land use/ land cover map prepared from landsat imagery (FCC) through visual interpretation.

## SECTION – III

## CHANGING ECONOMY

From the distant past the Paliyans were free from the monetized market economy. Their subsistence economy was based on foraging pursuits, and the few items that they required in the past were fulfilled through barter exchange. With the reservation of the forests and transformation of forest to coffee plantations, fruit orchards, or commercial crops, there was drastic destruction of forest ecology, which affected the traditional economic pursuits of the Paliyans. With the establishment of forest department by British, the Paliyan's activity came under the control of government regulation. The Paliyans were restricted from exploiting the forests freely. Thus with the changing ecology the traditional economy of the Paliyans was seriously affected. At this stage the arrival of the monetized market economy transformed traditional economy of the Paliyans.

### Changing ecology and its impact on the Paliyan culture

Large scale destruction, reservation and transformation of the forests seriously deprived the Paliyans to exploit the forest resources. The changing ecology brought two major changes in the Paliyan culture.

(a) Diminishing isolation of the Paliyans

(b) Changing Paliyan economy

### Diminishing isolation of the Paliyans

During colonial rule, for the first time, the physical isolation of the Paliyans was drastically affected. As discussed in previous section, with the construction of roads on high hills, establishment of urban center on high altitudes along with the growth of coffee plantation and fruit orchards, the physical isolation of the Paliyans was greatly reduced.

Construction of dams, roads, buildings and other developmental project led Paliyans to work with other labourers of the plains. In the day to day affairs they had to interact with police, revenue and government officials, health workers, forest contractors etc. The rural folk of different castes started working with the Paliyans in collection of fuel, cutting of timbers along with the collection of some of the non-timber forest produce. Traders from the nearby plains area came to the Paliyan hamlets to sell their goods and to purchase some of the forest produce (MFP). Thus through such close interaction, the isolation of Paliyans was substantially reduced with migration of large number of plains people to the Western Ghats.

## Changing economy of the Paliyans from foraging to labourer

Originally, the Paliyans were a nomadic foraging tribe surviving on roots and yams collected from the forest and hunting activities. Consequent to commercial utilization of forest products, urbanization of Palani hills, conversion of forest land into commercial plantation and agriculture by plains people and other migrants, the forest around the Palani hills faced erosion and so also the resources used by the Paliyans for their living. Even the little forest that remained in limited areas, from the above invasions did not serve as a free abode for the Paliyans any more. Various forest laws restricted them from free utilization and access. This changed the economic life pattern of the Paliyans.

Most Paliyans, baring a few, had no knowledge and expertise in agriculture since basically they were a nomadic tribe. Very few Paliyans owned some agricultural land in the upper valleys. But most did not have land since forest provided all their needs.

During colonial rule, the impact of industrial revolution was felt all over India and it affected the agriculture based peasant economy of Indian rural society. Such changes also affected slowly but effectively the gleaning economy of the Paliyans. They had the advantage of thorough knowledge of forest in Palani hills. With reservation and commodification of the forest the Paliyans were employed by the forest department as labourers to cut the timbers, to prune the trees and for the construction of forest roads. Subsequently with the growth of plantation substantial number of tribals worked as plantation labourers. Yet through out the colonial period the majority of the Paliyans depended on their foraging economy and continued to have barter in some areas. In earlier stage barter was in the form of silent trade. This further developed into direct exchange of goods.

As agricultural labourers under the agriculturists they were paid in kind. In most of the tribal and rural areas of India till independence the agricultural labourers used to work as bonded labourers. Paliyans most often preferred their foraging practices than to act as bonded labourer. But under the changed circumstances they had little option. Forest contractors used the practices to their advantage.

Gradually the cost of forest products, both the timbers and non-timber produce, increased drastically. The herbal items like myrobalam had international market, the price of tamarind and other minor forest produce used in dyes, soaps and cosmetic etc brought very good price. Paliyans worked for the forest contractors for collection of above items. Yet most of the profit of such items went to forest contractors and businessmen. Payment for these items were made in cash. Migration of large number of business men brought money economy to the forefront of Paliyans. Gradually they migrated to other areas to work as labourers.

In brief, in a slow process when forest became scarce the Paliyans were converted either as labourers or bonded labourers. They slowly adopted the work of labourer either along with their collection of minor forest produce or took up plantation work as a regular employee or as ordinary wage earning labourer of other advanced communities.

In Chapter – III we have discussed how the economic change in the Paliyan society took place through diversification of occupations.

In the traditional economic life of the Paliyans there were mainly two types of economic activities which consists of (1) foraging (a) hunting and fishing (b) food gathering (2) Collection of Minor forest produce.

But in course of time they adopted many new economic activities such as agricultural collection of MFP for contractors, labourer in plantation and field service etc. Dependency on foraging as a principal economic activity reduced to leisure sport.

In our sample, we have taken two villages – Palamalai as a developing village and Shenbagathope as a traditional village. At the same time both villages are selected in such a way that both of them are exposed to developmental processes. In chapter – III, we have put Palamalai at the 4th stage indicating that it is more developed than Shenbagathope which we have put between the third stage and 4th stage.

In Palamalai, majority of the Paliyans are labourers and they have limited opportunity for planning their future or for taking steps for their own development. They are well treated by the missionary 'Fathers' and the scope of exploitation is rather limited. Besides, for any developmental efforts one should have at least some resources, land, money or technical know-how. The missionary fathers pay the wages as per rules and takes care that the employees in the estate are not exploited or cheated. Besides, they make some labourers permanent and therefore they (labourers) enjoy some fringe benefits. Further the 'fathers' encourage the hard working labourers to adopt cultivation and guide them how to acquire Govt. lands through legal processes.

The diversified economic activities have been shown with two tables representing two villages. (Table No. 4.4 & 4.5). These two tables indicate

**Table No. 4.4 – Paliyans of Palamalai engaged in different type of work.**

| Number of House Holds | No. of memb ers | Number of working people | Number of dependan ts | Number of people engaged in service | | | | |
|---|---|---|---|---|---|---|---|---|
| | | | | In-sevice | Labour | Agriculture | Hunting & Gathering | MFP (Minor forest produce) |
| 1 | 6 | 5 | 1 | 1 | 4 | 4(s) | - | - |
| 2 | 2 | 2 | 0 | - | 2 | - | - | 1(s) |
| 3 | 4 | 2 | 2 | 1 | 1 | 1(s) | - | - |
| 4 | 7 | 4 | 3 | 1 | 2 | 2(s) | 1 | - |
| 5 | 3 | 2 | 1 | 1 | 1 | 1(s) | - | - |
| 6 | 7 | 2 | 5 | 1 | 1 | 2(s) | - | - |
| 7 | 7 | 5 | 2 | - | 5 | 2(s) | - | - |
| 8 | 2 | 2 | - | - | 2 | - | - | 1(s) |
| 9 | 4 | 2 | 2 | - | 2 | - | - | - |
| 10 | 5 | 2 | 3 | - | 1 | - | - | 1,1(s) |
| 11 | 7 | 3 | 4 | 2 | 1 | 1(s) | - | - |
| 12 | 10 | 4 | 6 | 1 | 3 | 3(s) | - | - |
| 13 | 4 | 2 | 2 | - | 1,1(s) | - | - | 1 |
| 14 | 9 | 6 | 3 | 2 | 4 | 1(s) | - | - |
| 15 | 3 | 2 | 1 | - | 2 | 1(s) | - | - |
| 16 | 6 | 2 | 4 | 1 | 1 | - | - | 2(s) |
| 17 | 3 | 2 | 1 | - | 2 | 1 | - | 1(s) |
| 18 | 2 | 2 | - | - | 2 | - | - | 2(s) |
| 19 | 5 | 3 | 2 | - | 3 | - | - | 2(s) |
| 20 | 5 | 3 | 2 | - | 3 | - | - | - |
| TOTAL | 101 | 57 (56.44%) | 44 (43.56%) | | | | | |

| | | | | | | |
|---|---|---|---|---|---|---|
| | Primary Occupation | 11 (19.30%) | 42 (73.68%) | 1 (1.76%) | 1 (1.76%) | 2 (3.5%) |
| | Secondary Occupation | - | 1 | 18 | - | 10 |

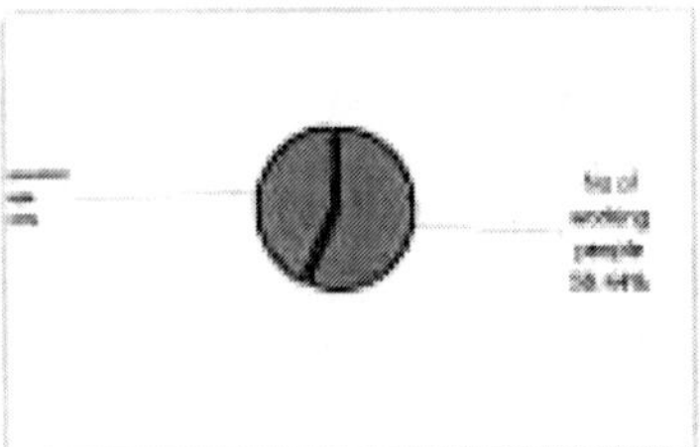

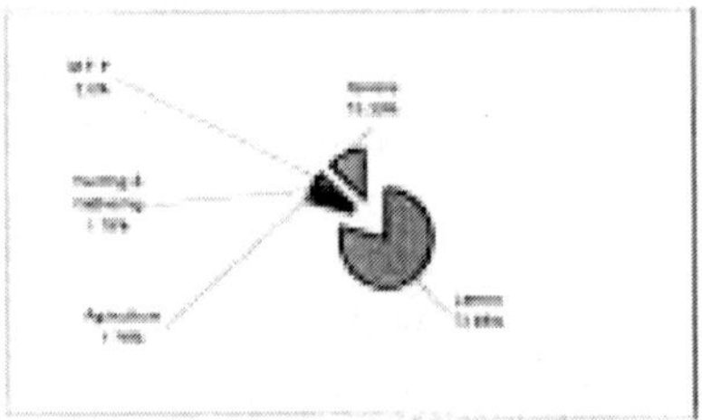

(s) refers to secondary occupations

**Table No. 4.5 – Paliyans of Shenbagathope engaged in different types of work.**

| House Hold No. | No. of Members | No. of working members | Dependents | No. of people engaged in | | | | |
|---|---|---|---|---|---|---|---|---|
| | | | | Service A | Labour B | Agriculture C | H&G D | MFP E |
| 1 | 4 | 3 | 1 | - | 1(s) | - | - | 3 |
| 2 | 5 | 1 | 4 | 1 | - | - | - | 2(s) |
| 3 | 2 | 1 | 1 | - | 1 | - | - | 1(s) |
| 4 | 3 | 2 | 1 | - | - | - | - | 2 |
| 5 | 4 | 3 | 1 | 1 | - | - | - | 2,1(s) |
| 6 | 3 | 3 | - | - | 1(s) | - | - | 3 |
| 7 | 2 | 2 | - | - | 2(s) | - | - | 2 |
| 8 | 2 | 1 | 1 | - | 1(s) | - | - | 1 |
| 9 | 2 | 2 | - | 1 | - | - | - | 1,1(s) |
| 10 | 4 | 3 | 1 | - | 3(s) | - | - | 3 |
| 11 | 2 | 2 | - | - | 1(s) | - | - | 2 |
| 12 | 2 | 2 | - | - | 2(s) | - | - | 2 |
| 13 | 2 | 2 | - | - | 1(s) | - | - | 2 |
| 14 | 2 | 2 | - | - | 1(s) | - | - | 2 |

*(Contd...)*

| 15 | 2 | 2 | - | - | 2(s) | - | - | 2 |
|---|---|---|---|---|---|---|---|---|
| 16 | 3 | 2 | 1 | - | 1(s) | - | - | 2 |
| 17 | 2 | 1 | 1 | - | - | - | *1(s) | **1 |
| 18 | 2 | 2 | - | 1 | - | - | - | 1 |
| TOTAL | 48 | 36 (75%) | 12 (25%) | | | | | |
| | | | | | | | | |
| | | Primary Occupation | | 4 (11.12%) | 1 (2.77%) | - | - | 31 (86.11%) |
| | | Secondary occupation | | - | 16 | - | 1(s) | 5(s) |
| | | (s) refers to secondary occupations.<br>**Mostly collects medicinal plants and herbs | | | | | | |

the distribution of workers and dependants. Further the working population of each village has been categorized under different occupations. With the help of pie chart the percentage of working force in each category of occupation / economic activities have been represented along with the dependent population. Economic activities have been categorized as primary and secondary in a separate table No. (4.6). the workers of both the villages are represented under primary and secondary occupation.

**Table No. 4.6 – Occupational details and composition of Paliyans at Palamalai and Shenbagathope**

| Types of Occupation | Palamalai | | Shenbagathope | |
|---|---|---|---|---|
| | Primary | Secondary | Primary | Secondary |
| Service | 11 | 0 | 4 | 0 |
| Labour * | 39 | 1 | 1 | 19 |
| Agriculture | 1 | 18 | 0 | 0 |
| H&G | 1 | 0 | 0 | 0 |
| M.F.P | 1 | 10 | 35 | 5 |
| Other | - | - | - | - |

* Labour includes in daily wages irrespective of place of working
* H & G even though practiced by few people does not appear in the table since no body acknowledges this because of Forest Act H & G – Hunting and Gathering
* Labour includes in daily wages irrespective of place of working*H & G even though practiced by few people does not appear in the table since no body acknowledges this because of Forest Act H & G – Hunting and Gathering.

Study of tables 4.4 and 4.5 reveal interesting informations about present Paliyan economic life. These are as follows:

## Demographic Pattern

In Palamalai 20 families interviewed consisted of 101 members and at Shenbagathope, 18 families consisted of 48 members. The latter indicating better alignment towards traditional Paliyan system, in which the family is smaller comparatively. The average family sizes are large at Palamalai (5.05) and small at Shenbagathope (2.66).

No. of dependents at Palamalai is greater (44 persons i.e 43.56%) than that at Shenbagathope (12 persons i.e 25%). The cause may be attributed to increasing education of children and care of senior citizens. On the contrary, families at Shenbagathope pursue economic activities with the domestic labour group.

## Occupation

The Paliyans at both the places appear to have moved away from hunting and gathering activities. Although the profession does not appear in table due to denial by the Paliyans, it is generally believed that the practice still exist. However, it is more in the form of part time or occasional allurement than a source of living.

Paliyans at Palamalai are more dependent on labour as their primary occupation. They have agriculture & M.F.P. collection as secondary occupation, which shows a trend towards monetized economy.

Paliyans at Shenbagathope, consistent with their traditional life, still followed MFP collection as their primary occupation and labour as their secondary occupation.

In both places women also participate in income generation and contribute to economic activity.

With the emergence of plantations in Palani hills, the estate owners were in need of labourers. They used to allure the Paliyans to work in their fruit orchards or coffee plantations. The older generations stuck to their traditional economic activities and could not adopt with changed pattern. But most Paliyans went to the plantations or orchards to earn their living. Those who could not adopt new opportunities had a miserable life. Those who had a little better financial situation due to adoption of new jobs were encouraged to work more for their food, drinks, entertainment, transport and medical expenses. Children who earlier lend helping hands in collection of M.F.P. now went to schools. Thus who had some better financial resources due to informal job opportunity had the privilege to spend their surplus money for a better life. This tendency is more common among the younger generation of Paliyans. And these people are inclined to adopt the changes compared to the senior members of the community. In Palamalai missionaries estate, workers are employed sympathetically and the valuation of work on every individual is not done rigidly. On the contrary in the case of other employers take the advantage of contractual labour agreement. For a piece of land (locally known to be a *Sadara*) based on the job involved, a rate is fixed. So, senior members sometimes do not fit to the trend due to their physical inability.

Paliyans in Palani hills also have exposure to various commodities at their doorsteps. In Palamalai there are two petty shops who sell all necessary items. The shop belongs to caste groups. Other than spending money on meals, Paliyans also spend in snacks, tea, bidi, cigarette, alcohol, film, clothing etc. Hawkers of both new and old garment came to their settlement to offer goods on installment basis.

The situation has changed a lot from the nomadic stage. With the flow of cash into the Paliyan life they are taking advantage of modern amenities. Now they are able to commute from one place to another by bus. They are able to visit hospital at far off places and are not hesitant to merge with mainstream. One Paliyan of Palamalai is able to take his horticultural products to the auction centre at the nearby town & get the return on prevalent market price. He has opened an account in the bank similar to others in the locality who have opened their account in Post office or involved themselves in private micro banking schemes. Due to changing ecology, their links to the surrounding forest has been drastically affected owing to reduction of forest area and forest regulations. Paliyans who led independent life in the past now became subservient to forest officials, forest contractors and become regular labourers in the plantations. They not only lost their carefree joyous life but they had to work hard to eke out their living and become labourer without earlier freedom.

In most of the plantations (owned by minority group private entrepreneurs) they are poorly paid by some measure of broken rice while women workers are reportedly exploited.

In the post independence era with the initiation and acceleration of developmental process, there has been tremendous change in the ecology of the Paliyan habitat in the Western Ghats. But since the objective was economic growth it also provided new job opportunities. The Paliyans had to alter their way of life. Previously they were living in a relatively isolated area. With the exodus of plains people to the hills, the Paliyans got scope to interact with large population of the incoming outsiders. Further with the growth of plantation, fruit orchards and cultivation of commercial crops they got the scope to work as labourers in the agricultural fields. At Palani hills and elsewhere from the beginning, the tribal people were exploited to a great extent. Their wages were paid in some measure of broken rice. Subsequently the Government passed abolition of bonded labourer act and minimum wages Act. It took a long time for the Paliyans to get information about such Acts. Nor did they know how to oppose against injustice. The immigrants who came from the plains were not necessarily richer segment of the society. But they were acquainted with monetized economy and habituated with the saving practices. With small capital, they tried to build up their fortunes by trying to carryout trade in these tribal pockets. With the availability of many varieties of consumer items, the food habits and recreation of the Paliyans changed. This process continued till recent past. But through this long period of transition they remained poor as usual though they were goaded by the forest contractors to work hard.

Thus within the last hundred years Paliyans underwent the transitional phase when most of them had to give up their foraging habits and settled down in permanent settlements. During the transitional phase the Paliyans were introduced and integrated into the monetized economy of the modern world, Those who still continued their age old foraging practices of traditional socio economic life, were forced to remain poor as usual.

Thus after being integrated within the monetized economy, the Paliynans by and large had to lead a poor life of labourer and had to struggle hard for their sustenance in a competitive world in spite of emergence of new opportunities which were available at their doors but being utilized by outsiders.

The Paliyans are slowly giving up their traditional way of life and trying to adopt new economic opportunities. But they are much exploited by the moneylenders, liquor vendors and above all by the forest contractors who purchased their forest produce at a negligible cost. Inspite of large-scale efforts by Government institutions and voluntary organizations they have remained illiterate, and led a life of poverty without reasonable status with the competitive life of the modern days.

It is interesting to note that in general after being labourers in plantations most of the Paliyans were exploited by the plantation owners. But in Palamalai the situation is some what different. The missionaries right from the beginning converted the Paliyans of Palamalai area and majority (or almost all) Paliyans who embraced Christianity, worked in missionary estate as plantation workers. As per the rules of the missionary estate some of the labourers were declared permanent and enjoyed service benefits such as free medical service, certain types of leaves and paid holidays etc. But more important to note is the fact that with guidance of missionary father some of these semi nomadic foraging people could own some land and could lead the life of settled cultivators. The table shows number of land owning Paliyan cultivators of Palamalai and the amount of land they owned (Table No. 3.6).

## Exploitation of the Paliyans in the Palani hills

We have discussed the situation in Palamalai where large scale plantation and agriculture has transformed the economic life of the Paliyans. But elsewhere in Palani hills the Paliyans could not obtain full benefits as they were heavily exploited by the plantation owners. There also, like Palamalai, there are advanced planters. Unfortunately, due to denied access by owners I could not get scope to make an investigation as regards the Paliyan labourers in those plantation estates. However a paper based on field investigation by Borian, a faculty member, Rural University Gandhigram is referred below to delineate the nature of exploitation in the plantation estates of the Palni Hills.

(1) **Encroachment of Paliyan land:** Majority of owners of coffee estates are high caste people or belong to the minority group. They were earlier struggling in the plains for their subsistence. But in Palani hills vast stretches of land were possessed by the tribals. The present plantation owners purchased such land for a song by offering "packets of sweets, pieces of clothes, palmgur cakes, idlies and dosas, bidies and cigarettes, betel leaves and tobacco and other trivial objects."

"In most cases, transfer of ownership was effected surreptitiously and the tribals were totally unaware of any such serious transactions."

(2) **Socio economic exploitation:** "The primary problem confronting the tribal plantation workers on this hill centres around economic exploitation, social subjugation and sexual harassment. The economic harassment was making them bonded labourers. (Borian: 1993 unpublished paper).

"Economic exploitation of these tribals is perpetrated in the form of payment of nominal wages, providing paltry advances which remain alive across the generation, not making the tribals as permanent workers and a host of other practices".

The plantation owners follow a multitude of strategies to evade the Government rules and exploit the workers. "Many estates are administratively fragmented into pieces of less than 5 hectares in order to be exempted from the preview of minimum wages act."

## Minimum Wages

By 1991 April when Tamilnadu Govt. had prescribed Rs. 15.28 per day as the minimum wages, most of the estates were paying only around Rs. 12/- for a male worker and Rs. 8/- for a woman as well as child workers on an average. On the lower side, a female worker was paid Rs. 6/- while a male worker is paid at Rs. 8/- in a few places.

Hundreds of large estates, employing thousands of workers had not made them permanent, although they had been working for several years with the same employer.

Besides, the Paliyans are paid lower wages as against labourers belonging to higher castes. Some estates pay the Paliyan tribals Rs. 10/- for male workers and Rs. 6/- for female workers as against Rs. 13/- and Rs. 8/- for non-tribal workers.

In these areas also, a few of the Paliyans work as collectors of minor forest produce (MFP) at the behest of authorized forest contractors. A local voluntary organization says that these tribals receive only marginally higher wages but their labour fetches the fantastic returns to the contactors.

Borian (Unpublished Paper) describes

"Paliyan tribals are supposed to be most hard working lots and the rich people and contractors are reported to make the best use of them, both for legitimate and illegitimate purposes. The potentials of these tribals are reported to be plundered for stealthily collection of sandalwood, brewing of illicit liquor, smuggling of timber and sale of ganja. As regards sexual harassment some elites maintain proximity with selected female workers, While in many places it goes with mutual consent for corporal pleasure, it does have an element of compulsion when it is used for commercial purposes. Perpetrated by a small set of bestial elites, this practices needs condemnation in order to protect gullible tribal girls their 'misuse'".

Finally Borian has emphasized that most of the Paliyans lead a poor life always remain indebted to their employers and thus they remain ever exposed to exploitation.

The preceding discussion about the condition of Paliyan labourers in Palani hills in general shows that they were exploited by the plantation owners and forest contractors and new opportunities of economic changes did not substantially improve their economic conditions. Condition of labourers in missionary estates in Palamalai is some what better.

If we compare the situation of the Paliyans at Palani hill with the paliyans of Srivilliputhur Taluk and adjoining areas we can see significant differences in the ecological set up and economic condition in these areas. In Palani hill the foraging habits of the Paliyans have been reduced substantially and they have almost given up their nomadic habits. The large number of estates having coffee Plantations, fruit orchards and other commercial crops have augmented the economic situation in this area. In the monetized economy they are in a better position. In general, they have improved their economic conditions as the commercial crops bring more profits and the payment of wages are to some extent influenced by government rules and regulations.

On the other hand the Paliyans of Srivilliputhur retain greater amount of self employment through collection of minor forest produce. But traders and contractors get the lion share of their profits the poor Paliyans remain in economically disadvantageous situation. Those Paliyans who work as agricultural labourers also do not get adequate wages as paddy and other crops are not very profitable as against cash crops and the wages remain comparatively low. The major source of income is through sale of MFP. But due to exploitation they are not able to substantially improve their economic condition. Some of them are indebted and till recent days some have situation akin to bonded labourers.

Paliyans in both the areas have lost their earlier economic independence due to deforestation and changing ecology and have adopted the life of labourer in most cases. In independent India, Govt. formulated schemes and projects for improvement of the tribal economy which we would discuss in the next chapter in the context of Paliyan's benefit.

## DISCUSSION

In the present chapter an effort has been made to show how, since the colonial rule, Govt. restricted exploitation of forests by tribals and rural people, with commercial motive. The commodification of forest created ecological hazards and also deprived rural people, more particularly tribal communities from utilization of forest resources for their sustenance. The ecological hazards along with monoculture forest, destroyed the rare biological specimens and depleted economic resources of the tribal communities including Paliyans.

With the growth of coffee plantations, and fruit orchards entrepreneurs migrated to the Western Ghat hill ranges and tribal people were forced to lead the life of labourers. The migration of large number of traders introduced various types of consumer goods and tribal communities remained poor due to exposure to these items, consumption and exploitation by traders. Due to the new monetized economy tribal communities, who had no saving habits, faced poverty due to poor wages and wasteful extravagant expenditure. Thus in spite of new economic opportunities tribal communities were heavily exploited and remained poor, illiterate and deprived of higher quality of life.

In the next chapter Govt. schemes and projects for improving the socio economic life of the tribals in general and the efforts for the development of Paliyans in particular has been presented with a view to understand the developmental process within the Paliyans society. In the analysis of development efforts, the factors that bring success and the circumstances that lead to failure will be highlighted. In brief, the main roots of the pauperization of the Paliyans can be attributed to changing ecology and advent of new monetized market system, coupled with migration of large number of people to the Paliyan habitat. Some of the migrants became entrepreneurs and with the exploitation of the new opportunities the migrants by establishing commercial plantations and fruit orchards became successful plantation owners. On the contrary Paliyans were employed with poor wages.

# 5

# Evolution of Tribal Administration and the Present Development Amongst the Paliyans

## SECTION – I

## TRIBAL DEVELOPMENT IN INDIA

### Tribal situation in India before the colonial rule

In the long history of Indian civilization, there are innumerable instances of close interaction between the tribes and the caste groups. The level of interaction spans over various facets of Indian society but had its limitation due to their isolation.

Though in the long history of Indian civilization there had been strife's between tribal's and more advanced non-tribal communities, the ruling Rajas used to retain cordial relationship with tribal chiefs and thus a smooth peaceful social relationship could be established. Sometimes powerful tribal chiefs could establish their authoritative political dominance and thus became part and parcel of advanced Indian society and the common tribal communities became integrated within the caste society.

Baring occasional strife's between ruling elites and common tribal communities there had been a peaceful co-existence between rulers and common tribal mass. No doubt on occasions, there were despotic rulers but tribal communities along with support of other castes could revolt and earlier mutual reciprocity could be re-established. By the time British established their authority over most of the tribal areas they were surprised by the great loyalty of tribal communities towards their Rajas.

During pre-colonial days the rulers of particular area could win over tribal communities and secured their loyalties through their cordiality and sweet personal relationship with the tribal leaders. Thus the Raja had to spend no money or effort in the administration of tribal areas where the tribal communities could establish their self-rule through their powerful popular leaders.

## Administration in tribal areas during colonial period

**Policy of segregation in pre-independence period:-** Britishers segregated the tribes and their areas from the general masses and developed separate administrative mechanism. Thus they were rightly credited with being the first to extend the arms of government most seriously to the inaccessible hilly terrain. The effort of such initiation can be tracked back to the time of Warren Hastings when he tried to bring the bandit "Hill men" or Paharias of Raj Mahal hills under the pale of civilization through the indefatigable Brooke, Browne and Cleveland in the later part of eighteenth century (Sinha S.P: 1981: 66). Military officers were preferred for governing tribal areas where there was revolt or unrest. They also introduced 'non-regulation'system and 'regulation system'. While the former was in tribal area and controlled by military officers, regulation system was controlled by civilians. But then there were lot of tribal revolts in 19th century which forced the Britishers to try various options. "Pax Britanica was meaning less without the religion of Christ being the major pursuation of the subject people & therefore a thorough & systematic induction of Christian missionaries started in the tribal area "(Sinha S.P. 1981: 68).

This isolation of tribals created by Britishers led to more exploitation by money lenders, traders & those who wielded power. Again land and forest, the twin needs of tribal people were under new regulatory mechanism. So common resources were brought under the control mechanism created by Britishers. The programme of systematic utilization of conservation and preservation of forest resulted in encroachment on the rights enjoyed by tribes for centuries (Elwin: 1963). As the colonial rulers had their eye on the commercial value of forest resources, they employed forest contractors who acted in the interest or Britishers. The emergence of this new species, in the forest life, who carried away the profit of forest produce resulted in exploitation of tribals.

It was not un-natural that these incursions into the tribal rights over forests resulted in sporadic retaliation which eventually led to the dislocation of the tribal economy & guarding the people from possible consequent frustration in the tribal mind. (Elwin: 1963:8).

In the tribal areas infrastructural developments such as railways, road etc. developed only where Britishers had their economic interests. So new roads were constructed in the tribal areas for transportation. The policy of not developing the tribal areas led to a sort of feeling of isolation and inferiority complex among the tribals. The isolation began with the enactment of Government of India Act of 1870.

The colonial Government enacted the Scheduled District Acts in 1874 and kept large tracts of tribal areas outside the jurisdiction of normal administration.

A few tracts were specified as scheduled areas. In course of time many acts such as Scheduled District Act, Backward Tracts Act etc. were enforced from time to time till 1919. The Backward Tracts Act were administered on the suggestion of Montegue and Chelmsford on the basis that these people were primitive without political institution. Further, as per 1935 Act, two categories of tribal areas, i.e. "excluded areas" & "Partially excluded areas" were created under section 91 and 92 in 1936. The main features to distinguish an excluded area from a partially excluded area were:

(i) Governor acted in his own discretion in an excluded area where as he sought the permission of the ministers in a partially excluded area.

(ii) The expenditure in the case of excluded area was non – votable while in the case of partially excluded areas were subjected to a vote of the legislature.

(iii) And any debate or discussion on excluded area needed prior permission of Governor. This exclusion served to a great extent the British purpose of keeping the tribal area isolated from the main stream of national life. The restrictions imposed by the administration in "excluded areas" and "partially excluded areas" aimed at checking influx of non-tribals to tribals areas in order to prevent tribal exploitation by the outsiders.

After independence Jawaharlal Nehru, the first Prime Minister of India, had declared that "We should help the tribal people to develop along the lines of their own tradition and genius, teaching them not to despise their past, but to built upon it" (Quoted in Elwin, 1963:5)

## Tribal development in post Independence Era

The great leaders of our country who gave us the independence also had thought about the different sections of our people. Necessary Provisions were made in the Constitution of India for safeguarding rights and interests of Scheduled castes and Scheduled Tribes who constitute more than 22% of total population provision for safeguard can be traced from the very first session of Constituent Assembly. In December 1946, Jawaharlal Nehru had outlined: "adequate safeguards shall be provided for minorities, backward and tribal areas and depressed & other back ward classes". Article 46 under Directive Principles of State Policies in Part IV of the constitution provides:

The state shall promote with special care the educational and economic interests of the weaker sections of the people, and in particular of Scheduled Castes & Scheduled Tribes and shall protect them from social injustice and all forms of exploitation. *(Elwin: 1963: 16).*

Special safeguards have been provided in the Constitution of India & have been spelt out in 24 articles and in Fifth and Sixth Schedules, which can be referred in appendix of the constitution of India.

After the independence of India it was a great challenge for the national leaders to build a prosperous modern nation as postulated in the Indian constitution. When Nehru announced on 14th / 15th August 1947 the historic statement "India's tryst with destiny" it was sure that he thought of five lakh Indian villages and about the aborigines of India. So it is essential to see the tribal development in the context of general development in the country. The Governmental effort in bringing about rural development were implemented either directly or indirectly by various government departments independently. The visible absence of team work and co-ordination resulted in misdirection of efforts and wasteful expenditures. (Dube 1982:22). To do away with such short comings the government was determined to change the policy of rural development during this period The concrete design for rural development and the financial assistance needed to implement it came from United States of America. The first five year plan was started in the year 1951.

The community development programme was started on 2nd October 1952. These programmes covered a wide spectrum of activities in agriculture, animal husbandry, education, health, communication, social services, rural industries, panchayats and so on. For the tribal areas, it was decided to supplement the community development programmes as these areas comprise of hilly and forest regions and are sparsely populated with little institutional infrastructure. Accordingly forty-three special Multi purpose Tribal Development Projects [M.T.D.P] were started during 1954. During this first five year plan though money was spent on (a) Education (b) Economic upliftment (c) Health, Housing & Communication etc. yet the emphasis was on Health, Housing and Communication etc. In the Second Five Year Plan the emphasis shifted to economic development. Here areas such as agriculture, cottage industry, forest Co-operative societies & setting up of Multipurpose Tribal Blocks were given importance (Sinha 1981). In the Third Five Year Plan, these blocks were renamed as Tribal Development Block (T.D.B) to cover more tribal areas consisting of 150-200 square miles & a population of 25000. The focus in this Third Five Year Plan was on Economic Upliftment, Education & Health, Housing & Communication. Thus by and large, areas with two third tribal population only got the benefit of this special development effort up to the end of the fourth five year plan (Sharma B.D.).

As these programmes were land based it was decided to cover small land holders and landless people. Attention shifted from area development to development of identified families and in the Fourth Five Year Plan, a series of programmes were conceived for rural areas, namely, (a) The small farmers Development Agencies (b) Marginal Farmers & Agricultural labourers Development Agencies. Thus, the trend was to give attention to weaker sections. In the above ideology some tribal areas experiencing tribal unrest was taken up. Six Tribal Development Agencies (T.D.A.) were established on pattern of S.F.D.A. with higher investment during Fourth Plan. The T.D.A.s were expected to comprise elements of economic development, social services and protective measures. But in reality emphasis was on agriculture & roads.

The task of tribal development was not stereotyped planning & implementation. The heterogeneous problems in the tribal society made planners to ponder over new strategies.

The first review of the special Multipurpose Tribal Development Projects was takenup by Elwin Committee and an all inclusive review of tribal scene was done by Dhevar Commission in 1961 under Article 340 of the constitution. They noted (a) slow pace of development (b) meager investment in tribal areas impeding tribal development.

They also suggested protection to all tribals living within the tribal areas and outside and then extension of tribal development blocks with more than 50% tribal population.

Shilu Ao appointed in 1969 for reviewing the tribal development program accepted Dhevar's recommendation & felt that the main problems were related to indebtedness, land alienation, educational backwardness & inadequacy of communication. To overcome the above they suggested implementation of the recommendations of the commission and allocation of larger resources proportionate to the problems.

As each area presented a unique situation in terms of its resources, peoples, social constraints, and stage of infrastructural developments etc., the earlier approach had to be replaced by specific programme for each area. In 5th Five Year Plan the tribal development was broadly divided into two categories.

(a) Development of tribals living in areas of their concentration.

(b) Development of dispersed tribal population.

Now it had to be identified which tribal area should fall under each of the above category. Areas of tribal concentration were included in the 5th and 6th schedule of constitution. But there were certain discrepancies in the inclusion and exclusion of areas in the 5th and 6th schedule. So it was decided that, all areas with more than 50% tribal population (as recommended by Dhevar) were to be included under a special programme known as 'tribal sub-plan'. Its objective was to reduce the gap between levels of development of tribal and non-tribal areas and improvement of tribal life. Again another plan named 'minimum needs programme' (M.N.P) was launched in 5th plan period in order to provide basic necessities to the tribal people which included.

(i) Provision of elementary education.

(ii) Health services

(iii) Water supply.

(iv) Construction of roads.

(v) Rural electrification.

(vi) Provision of house sites for landless.

(vii) Launching of nutrition programme

(viii) Improvement of environment.

But it didn't cover their basic need that is food or their subsistence pattern. Again the above things were not the demand of beneficiaries (Behura: 1995: 108).

The task force constituted in 1972 for development of tribal areas suggested that the ecological, occupational and social parameters of the tribal population should be properly assessed for formulation of a policy and its implementation.

Shilu Ao Committee observed that the tribal development block was not a suitable unit for planning and implementation for the tribal areas because of its small size and therefore a new unit at the meso level was envisaged. Thus a group of blocks were selected by the respective states, while keeping in view factors like geopraphical condition and level of socio-economic development of people, constituted an Integrated Tribal Development Project (I.T.D.P).

For development of dispersed tribal population it was decided to take the group of family as unit. It was also realized that some communities needed care both at the stage of planning and also implementation of developmental programmes.

So in the 5th plan four parameters were recognized for tribal development particularly that of primitive tribes,

(1) Size of community.

(2) The stage of their economy.

(3) The extent of isolation.

(4) The level of literacy.

Size is important as larger one can adopt in extreme adverse condition while in the smaller one a small disturbance in total system may destabilize them.

Those communities which still subsist at food gathering and hunting stage qualified to be in the category of primitive tribals. The dividing line is the transition to agriculture as the main source of livelihood. All communities having less than 5% literacy were treated as preliterate groups and categorized as primitive. Nearly 74 S.T. communities have been identified till 1st June 1989 as primitive and separate micro projects have been identified for each of them (Behura N.K.: 1995:110). As the limit of each micro project is 500 families sometime two projects runs for larger primitive tribes. The Government of India funded all the micro project in the form of SCA (Special Central Assistance) on cent per cent basis through various State Government since 1975-76. The working group of 6th plan had suggested that (1) there should be unique programme for each group, (2) conservation and re-organisation of traditional scheme (3) and the development modalities and sectoral programme should consider the eco-system.

In the Sixth Five Year Plan, these smaller pockets of tribal concentration not covered in sub-plan were under an intensive integrated development on a Modified Area Development Approach (MADA) under tribal sub-plan. Tribal sub-plan emphasized on raising productivity level, eliminating exploitation of tribals, adequate infrastructure and to alleviate the problems of the ST families who are below poverty line. In the Seventh Five Year Plan the objective was to have a judicious mix of area cum family development while paying attention to the vulnerable section of ST. To have cent per cent financial support of tribals under TSP (Tribal Sub Plan) approach, areas having 5000 population having 50% or more tribals were considered. Among the important contribution of the tribal sub plan strategy over a period of years in our country is the gradual evaluation of the concept of the family oriented programmes for eradication of poverty in the tribal areas.

As has been discussed till now various tribal programmes have been conceived from the day of independence for the development of rural masses in general and tribal people in particular. Though efforts are made to recognize the smallest unit of planning, yet the actual assessment is done with an individual. But due to various practical problems one has to be content with group of people instead of the individual. So after various trials or experimentation Micro Planning and Growth Center concept emerged. "Micro Planning as a concept envisages identification of a small viable area for planning purposes, preferable with a homo geneous socio-economic landscape——The growth center technique can be adopted usefully while preparing micro level plans for providing a variety of complemental services at selected points in the area". (Sharma B. D. 1984: 154).

The techniques of growth centre theory was developed for planning purposes & it is not an essential part of micro level planning. But those villages which assume central location and strategic situation from the point of view of resource availability and existence of support services to cushion the new development programmes can serve better as growth centres (Sharma B.D.–1984). In Micro-level planning, the central theme is the comprehensive view of the total developmental matrix which may include many items but may not be strictly developmental in character like protective measures and administrative infrastructure in the case of tribal development.

In view of the rural urban continuum the urban centres in the advanced areas provide a base for diversification of the economy of a small region and acts as points of attraction for the population their in. This presents an idealized dynamic situation which is accepted as starting point for application of the growth centre approach in preparing a plan for balanced development of micro areas in a region. *(Sharma B.D 1984).*

But, in a plan, the programme package will depend largely on the growth model adopted by policy makers for development. However there are certain presumptions as elaborated by Sharma (1983).

(i) The growth stimuli have an even geographical spread effect.

(ii) The benefit of development percolate to the lower strata in the society in course of time.

(iii) The socio-economic topography of the region is even, which, by implication means that the spread effect and the percolation effect will automatically set in and it is only a matter of time for the benefits to spread out both socially and geographically.

(iv) Man power is an ordinary factor of production just like any other factor. Demand for different skills can be met from larger regions with no implication for the local community.

But it is not as simple as mentioned above since the level of socio-economic development of different sections of the population in these areas differs considerably. Sharma (1984) writes "it is clear that the growth center theory both in terms of its concept and its application through techniques developed for advanced areas is not suitable for the simple socio-economic situations of tribal area."

Although Government had modified the strategy of tribal development from time to time with a view to make the planning more effective and adopted greater intellectual concepts and approaches, they are also conscious of the great importance of financial support to implement such plans and modified strategies. A brief discussion on financial allocation will help in understanding the process.

## Financial Allocation

At this stage it is also necessary to know how the financial allocation, a vital ingredient to development, took place in post independence era. India's Constitution envisages independent sources of revenue for the states and union territories. Initially Finance Commission was set up to see the needs of states and on this basis union government gives the funds under Article 275 (I). Subsequently by Third Five Year Plan, Planning Commission became responsible for the allocation of funds for plan programmes while the Finance Commission was responsible for the remaining items. Briefly saying the financial resources for developmental programmes in a state may comprise the following elements.

(i) Investment in the Central and centrally Sponsored Schemes.

(ii) State Revenues.

(iii) Share from Central Revenues.

(iv) Plan assistance from the Central Government under Gadgil Formula.

(v) Grants under Article 275 (J) on the basis of the recommendation of the Finance Commission.

Generally tribal majority areas fall under three categories.

(i) Predominantly tribal states / Union Territory

(ii) Scheduled areas

(iii) Non-Scheduled areas

The consensus for financial provision was made by the Constituent Assembly as they had accepted tribal affair to be the national issue and above party considerations even though the tribal areas are integral parts of the states.

The first provision in Article 275 (I) is quite comprehensive and covers both developmental and non-developmental needs of the tribal areas and the tribal communities. The outlays under this provision are given as grant and charged to the consolidated fund of India. It is note-worthy that this proviso cuts across the normal financial relationship between the Union and States, envisaged in the constitution. So when the 2nd Finance Commission sought clarification about assessing the needs of tribal areas, Union Government replied that matters relating to tribal areas and tribal communities could not await a five yearly review of the Finance Commission, and therefore should be deemed outside their preview.

Assistance for tribal development during the two plan periods was made by Centre to respective states for their state plan under Article 275 (I) as cent per cent grant. Subsequently under a separate head that is “Welfare of Back Ward Classes (W.B.C.)”, centre and state provided matching contribution towards tribal development. With the introduction of Gadgil Formula, this grant formed part of the block grant and was shown under welfare of backward classes. This fund was put under three broad categories, (a) Education (b) Economic Development (c) Health, Housing etc.

Apart from this, additional assistance was provided under centrally sponsored Programmes, under the head-”Welfare Backward Classes

(W.B.C.)", Viz. Special Multipurpose Tribal Development Project, Tribal Development Blocks and Tribal Development Agencies, post matric scholarship, girl's hostel, cooperation, training and research etc.

## DISCUSSION

The whole planning process and its successive developments are great intellectual exercise regarding which common people has least participation at the planning stage. Plans are drawn off by economists, intellectuals and administrative bureaucrats majority of whom have no touch with people and their problems. For such planning, main aim is to implement these through grass root level institutions, with the co-operation of rural people. Majority of the people have no idea as regards what is welfare state and what should be their role in translating the plans and the projects into practical achievements. Nor the tribal people have the capital to invest and their co-operation could be generated through planned organization.

Major role of the administrators was to implement the projects by sanctioning the funds and to carry out the work through contractors in the tradition of British rule. Neither the people had awareness of developmental projects nor they had the national spirit to achieve development for the good of the common masses and especially weaker sections.

Besides, since the administrators had no close contact with the people, they could not successfully implement the projects. A few comments are given below as regards implementation of plan projects.

(a) People's participation both at the stage of planning and framing the projects and at the level of implementing these projects are seriously lacking. In fact beneficiaries or people in general, have very little scope to participate and cooperate in these developmental activities.

(b) In fact, as mentioned earlier, the five year plan projects are serious intellectual exercise by elite's and officials who have no touch with rural and tribal people.

## SECTION – II

## DEVELOPMENT AMONGST THE PALIYANS

In the preceding section, the plants and strategies of tribal development at the national level have been briefly presented. To analyse and evaluate to what extent these policies have been implemented is beyond the scope of the present study.

Moreover, to arrive at the exact impact of developmental process amongst the Paliyans in whole of Tamilnadu is also a difficult affair. Yet to focus the field situation and to evaluate the actual operation of these strategies at micro level, we have taken up both Palamalai and Shenbagathope villages and in special cases some of the near by villages.

The few projects that have been implemented in two villages or in their nearby villages will be discussed in this section so as to give a picture of the developmental process of Paliyans. Most of the developmental strategies in India emphasize on agricultural developments, as 70% of Indias population belong to rural sector. Traditionally, the Paliyans are food gatherers and most of them do not possess any land for cultivation. In such a situation government plans for poverty alleviation is not available to majority of land less Paliyans.

Government categorized certain foraging tribal groups as primitive tribes at a latter stage in 5th Five-Year plan. Accordingly the communities which still subsist at food gathering and hunting stage qualify to be categorized as 'Primitive tribes'. During the fifth five-year plan fifty two tribal groups were qualified as 'Primitive tribes'. The Paliyans could not be included in the above mentioned list. But in subsequent stage, during sixth plan, the tribe was recommended to be included within such list. This inclusion has a bearing on resource allocation to the Paliyans as per policy of Government. But till the completion of my field, work, it did not happen. However the plans and projects that have been implemented in the two villages or nearby villages which relates to the development of Paliyans is being discussed below.

The field study has been taken up on prominent developmental schemes, such as Housing, Health care service, Educational development etc. Which are vital sectors for sustained development.

## Housing Scheme

"For the first time in Indian history, a concerted national action to develop the non-sedentary tribal groups pratising collectional, hunting, primitive industry, slash and burn type of primitive cultivation, was taken up by the Government of India under the second five year plan as a public policy, though such individual actions were contemplated since the beginning of the British dominations in India, e.g. with the Paharias of Raj Mahal Hill and Baigas of Central India." (Malhotra: 1968:369). Since second five year plan such rehabilitation experiments started in different states.

In chapter, I, we have discussed the nomadic life pattern of the Paliyans and their temporary shelters. Accordingly, Tamilnadu Government with the intention of rehabilitating the Paliyan community, encouraged Government supported Group Housing scheme in Shenbagathope, which are in operation for the last four decades. Since the Paliyans lead a seminomadic life, housing scheme is considered most vital for their development.

## Group housing scheme at Shenbagathope

According to Headmaster of Upper Primary school the first housing scheme at the Shenbagathope is reported to tbe implemented in early sixties (1962) during tenure of Kamarajar as the Chief Minister of Tamil Nadu. Under this scheme, 13 houses, a school building with a kitchen for cooking the midday meals were constructed by Government. But only five dilapidated buildings are existing at present. The subsequent houses were taken up for construction under J.R.Y. (Jawahar Rojagar Yojana) in the eighties. Twelve houses were constructed in the second phase and fourteen in the third phase under this scheme. In addition, few utilities buildings were also constructed after third phase of construction. Of them, the construction of a large Kalyana Mandapam and a building for Co-operative society is note worthy.

## Pattern of houses introduced by the Government:

The houses are pucca buildings with tiled roofing and raised foundation. Each house consists of a single room with a low partition inside for separating a cooking area from the living room. At the back of the house, a toilet was constructed. But since the Paliyans are not acquainted with the toilet they have filled it up with sand. There is a large well on the main street adjacent to the group houses for portable water.

## Emerging settlement pattern

The settlement pattern at Shenbagathope is very simple. As one enters into the village one can see huses distributed in the right hand side in a single row. In the left hand side houses are distributed on either side of parallel lanes. The first phase houses which were constructed in the year 1962 are on the right hand side the residential school and the kitchen are in the left hand side. Next starts, the second phase houses which were started in early eighties while third phase houses are situated towards the end of the street. The quarter for the teacher, which was used as a dispensary and a store room for the medicines, is in the right hand side. Besides, the cooperative store and office are in the left hand side. The Kalyana Mandapam is in the vacant space behind the left hand side houses. This building was built with the financial support from the voluntary organisation (SOCO), which is closely associated with developmental activities in the Shenbagathope village.

At present, first phase houses are completely destroyed and no Paliyan lives there. The school building is fairly in good condition as it is repaired at intervals. The first phase houses were destroyed chiefly because most of the allotees did not take much interest in occupying and maintaining these houses.

During early eighties, group hosing scheme was implemented vigorously. The basic objective of this programme was to provide houses to houseless people. The programme comes under Jawahar Rojgar Yojana. Under this scheme the S.C. and S.T. families get the houses almost free. The second phase houses consisting of 12 buildings were constructed, and given to 12 beneficiaries. In the third phase, 14 houses were constructed and were allotted to equal no of households.

During my field work, I found that excepting two, all other houses of second phase were occupied. But all the third phase houses were lying vacant excepting three, which were occupied by Paliyans. These houses are in severely dilapidated condition while second phase houses are fairly in good condition.

The third phase houses were constructed hurriedly under pressure from the voluntary organization. The SOCO workers were successful in congregating Paliyans from the nearby hamlets with a view to build up a central place for their development. With that objective in view, SOCO renamed Shenbagabathope as 'Bhagavati Nagar' and tried to introduce developmental projects for the welfare of but, that did not succeed in fulfilling the desired goal. Paliyans who came from outside villages soon deserted Shenbagathope. For example, the inhabitants of Aathi Kovil who came to Shenbagathope faced difficulty in staying there. Local inhabitants and migrated Paliyans collected M.F.P. from the same area which led to conflict and clashes. So after two years, the people of Aathi Kovil left Shenbagathope and went off to their village. Other people left Shenbagathope in the same manner. Thus making this place a prosperous centre for development failed. The first phase and third phase housing projects have failed miserably while second phase housing scheme faced partial or small scale failure. The reasons for such failures can be explored. But prior to that we can analyse total settlement pattern of Shenbagathope including the scattered settlements of all people who claim to belong to this village. These people are united through various social, economic, religious and political network and have the sense of identity of belonging to the village Shenbagathope.

## Analysis of settlements at Shenbagathope

At present there are two types of residential settlements in the village Shenbagathope.

1. The traditional settlements which lie far away from the core village.
2. The core village which consists of permanent pucca buildings constructed under Government Housing Schemes.

The first category of settlements consist of age old living arrangements or shelters that they developed from the past traditional patterns by making temporary huts, using the locally available materials. These huts or houses are not durable for long period and require frequent repairs. But people prefer such houses as the construction and maintenance of these houses require very small amount of money inspite of distance from the core village. Besides, from these settlements it is convenient to carry out the traditional economic pursuits of hunting and collecting MFP.

In Shenbagathope there are two such settlements at a distance from the core village which are closely linked with their economic life. The first peripheral settlement is now defunct, but in the estate of the Andal temple there was a settlement previously, which is now abandoned after the death of a person. This defunct settlement sometimes is used as temporary shelter during collection of mangoes or other fruits. The second settlement is still continuing near the periphery of the guava orchard on the hill slopes nearly two kilometers away from the core village. This settlement is really the hub of the traditional Paliyan economic life. M.F.P.s, after collections, are stored there (some items are also dried) before transportation tot the forest contractor or to the local market. Since this site is very near to the area of economic operation of collecting M.F.P., it is a convenient place to follow traditional economic pursuits of 202 collecting M.F.P. Besides, this settlement provides the isolated and peaceful social existence for the Paliyans.

## Cultural barrier for adoption of colony houses / Permanent houses

The description of above settlement pattern in Shenbagathope gives us an insight to analyse the cultural barriers in adopting *pucca* buildings provided by the Government under group housing project. We now discuss the cultural factors that inhibit the adoption of such buildings which are much more durable than there traditional houses.

## Economic factor

In the village Shenbagathoppe, the Paliyans are invariably under economic dependency on the contractor of MFP till today even though the forest comes under the sanctuary area. As shown above, the Paliyans prefer to live in 'Medu' instead of living in Pucca buildings provided by the Government mainly because 'Medu' is nearer to the forest, which are sources for the collection of non-timber forest produce. Collection of MFP depends on various factors. Mostly these items are collected depending on the demand of the market. For example, a specific region yields particular item. Living nearer to the forest, it is easier to identify such locations, which contain the desired items having ready demand in the market. It is also easier to collect head loads of fagot to "Medu" for cooking purposes than carrying such head loads for 2 kms. Besides, 'Medu' provides access to other valuable items which ae usually prohibited by the forest department. There are other social factors for which Paliyans prefer to stay in huts in 'Medu' rather than living in Pucca houses, constructed by the Government under Jawahar Rojagar Yojana.

## Social and Cultural factors

The Paliyans not only desire to stay in isolated villages or hamlets but they also prefer to stay in small groups. Usually the Paliyan hamlets of Kamarajar district in further southern areas, consist of 8 to 10 families each.

The Paliyans are considered usually non-aggressive (Gardener). But this relates to their interaction with outsiders. Within the community when they live together there is constant fighting. In a Paliyan village, in the evening, when they are in a drunken state people scold their opponents freely with loud voice and abusive language. Yet they try to avoid such situation to occur, by living in a small number with most close kin members with whom they have sweet relationship. When SOCO workers made efforts to collect Paliyans from different hamlets (with or without close relationship) such conglomeration of Paliyans created social problems beyond solution.

When they were staying in small groups scattered in different areas, there was minimal conflicts. But once they started living together it gave rise to conflict within the community.

For the inhabitants of Aathi Kovil, living here was difficult as the area of operation (for collection of MFP) clashed with that of the local inhabitants. Even families who are part of Shenbagathope census but staying at 'Medu' (at about 2km away) for whom houses were allotted here left this place due to inconvenience in their collection of MFP. As a semi-nomadic foraging community, Paliyans sometimes change their homes and stay in their relation's house on various occasions. This has led to the situation for deserting the area by some persons of the village Shenbagathope.

Apart from bickering in the late evening Paliyan males fear family break up. This occurs when the wife is attracted towards another male within the community due to free mixing. Once this happens the situation becomes further complicated. A few cases (at least three) took place during my stay.

Paliyans have another custom of leaving a place if some body dies. The Paliyans had also the habit of leaving a place when they were affected by diseases like cholera, small pox etc.

The Paliyans usually avoid out siders. When they deserted the places for collecting MFP, outsiders occupied the vacant houses. On their return the Paliyans avoid conflict and leave the place consistent with their nomadic habits. In the first phase houses it seems such situations occurred.

Thus due to these socio-cultural factors, in addition to the economic factors, the Paliyans tried to avoid living in Government constructed houses at the initial phase, Subsequently they became adapted to such situation and made friendship with outsiders and resided in their allotted houses without fear or reluctance.

At present 2-3 families shifted from "Medu" to the core village based on their desire and felt needs.

The effort of the voluntary organization engaged in improving the Paliyans had some undesirable results. They tried to concentrate the Paliyans in a tiny village. This resulted in conflict for exploitation of economic resources as discussed earlier. Psychological incompatibility and differing behavioural values are the other factors responsible for desertion.

## Houses constructed by N.G.O

In the context of our discussion on Paliyan settlement, it is worth while mentioning the houses constructed by the NGO (ASSIFA) nearer to the sampled village area. The location of the area is at some distance from Sriviliputhur (Pilavakkal dam area). The houses were constructed for SC & ST and out of these six houses were given to Paliyans as displaced persons from the submerged reservoir. These houses are sturdy and comparatively of superior construction than the Govt. group houses.

The rehabilitated Paliyans seemed to enjoy the new settlement and the scheme appeared to be successful. The Paliyans were comfortable with the nearby forest surroundings which provides them with the facility of collecting MFP. One Dhevar man has married a Paliyan lady and the relationship between members of the community was very much cordial.

One remarkable contrast can be observed that while Paliyans at Shenbagathope are reluctant to occupy group houses constructed by Government, the Paliyans at Pilavakkal dam are attracted to it.

Perhaps less conflict because of small group and availability of M.F.P. in the nearby forest are the reasons behind such peaceful behaviour of the Paliyans at Pillavakal dam area.

## Group houses in Palani hills

The most conspicuous development programs in the Palani hills is the group housing scheme. There are over a half dozen group housing colonies at Kumapatti, Pala Malai, Vahragara Parai, Moolaiyar, Kadasikadu etc. constructed by both private and government agencies. We shall discuss about the group housing in Palani hills in general followed by elaborate description of one such scheme in Palamalai village.

The group houses constructed by the private agency (NGO) which constructed at Vahragaraparai seems to be more successful compared to Government constructed group houses. In the case of former houses the houses are spacious and well-built. The fund for the same was obtained from a foreign donor country, as the N.G.O. which was instrumental in construction of these houses belong to a missionary related organization. Even after few years of construction, there was no complain about roof leakage or any technical problems. Again the selection of the site was appreciated by Paliyans inhabiting there.

Since majority of the group houses in Palani Hills are sponsored by Government we have selected the group house at Palamalai for detailed study. As mentioned earlier the American Missionaries of Madurai in their search for a suitable site for a sanitorium for their staff landed at Palani hills (Chapter – II). Slowly others came and Jesuit fathers at Palamalai established their church as early as 1900 and in the same year converted 25 Paliyans to Christianity. They further established an extensive estate (approximately 250 acres) for coffee plantation and fruit orchards for which they recruited & converted Paliyans as labourers. For these labourers they constructed a housing colony for their residence. Thus Paliyans at Palamalai had the first experience of living in housing colony as early as the beginning of twentieth century which provided scope to live together with people of different villages including even Puliyans who belonged to scheduled castes. This first housing colony in a way set the model for living together by people of different villages.

The Government sponsored housing colony at Palamalai which emerged under Group Housing Scheme in 1986-87 is set up on the main road not far from the labour colony of the Palamalai estate. The houses are well built with RC roofs and each family was allotted one room only.

The selection of the allotees was made by the Government officials. As per rules of the Group Housing Scheme, only landless Schedule Tribe and Schedule Caste are allowed to get such houses. In the project 20 houses were fixed for 20 allotees. But in fact 21 houses were constructed and the extra one was allotted to a higher caste man. According to the statement of the occupant, he had paid a bribe of Rs. 2000/- and could obtain the allotment from the Government Officials. The NGO working in the area have wedged a war against such illegal work and they are demanding that the extra house can be converted into a club house or a guest house. This situation has resulted in some amount of conflict between tribals and non-tribals.

Again though 20 houses were allotted to 20 tribals it was revealed that two non-tribals were found occupying two houses which were allotted to two tribals. One of them is now living in the estate quarter which has three rooms and one of his nontribal friend is at present occupying his allotted house on payment of Rs. 1500/-. It is agreed that when the allotees retire from the estate work the non-tribal occupant will surrender the house after receiving his deposited money of Rs. 1500/- In another case the real allotee is the eldest brother of the previous allotee referred above. He has also given his allotted house to some non-tribal with the same type of agreement receiving Rs. 1500/ - as caution money with a condition to get back his house after returning the fixed amount of Rs. 1500/-. He in fact prefers to stay in the farm house of the joint family's agricultural field which he cultivates for the family. Besides looking after the fields the area is nearer to the forest where he hunts frequently as a hobby.

This case indicates that both the brothers are not land less but we are not sure whether agricultural land is Patta land or merely *Porombok* land.

In spite of such irregularities the Palamalai group housing project was very successful so much so that the tribal people who could not get such benefits were keenly interested to have a house under group housing scheme. There was such a great pressure from the deprived Paliyans ,that a local N.G.O. advised such people to take steps to posses a house. They were asked to select a suitable plot and stay there so that the N.G.O. could bring it to the notice of government authority and thus they would have their houses. The cook or attendant attached to the missionary house spared his land to be occupied by these people so that another hamlet would come up shortly. But slowly the N.G.O. who was working in this area showed little interest.

The night school they had opened for the children and elderly people also got closed. The cook who had initially volunteered to spare his land felt uncomfortable. His hope of getting back the land if the group houses are constructed in other land or getting compensation by giving his own land became grim. So during my last visit Paliyans inhabiting in this hamlet showed me how they were planning to go to a new site not far off from this existing one. They contributed this to the demand of the cook.

The above picture shows that the group-housing scheme is becoming popular in Palini hill area though it took a long time for the people Shenbagathope to adapt to it. The reason for such situation will be discussed below.

Though both Shenbagathope and Palamalai villages are inhabited by the Paliyans each has its own distinctive ecological factors, historical development and the process of acculturation.

Shenbagathope is at the foot of the hill and the villagers used to depend fully on diascoria and other items collected from the forest through foraging. The people of Palamalai after the establishment of church mostly acted as labourers in the estate and lost the habit of foraging excepting occasional hunting. Thus the people of Shenbagathope continued their seminomadic life almost till the present days while people of Palamalai adopted sedentary life much earlier on the high altitude of Paliyan hills. The people of Shenbagathope depended more on nature and could build their houses with the locally available material. The Paliyans of Palamalai were loosing these traditions very rapidly.

The Shenbagathope people had comparatively less opportunity for adoption of innovation of more durable houses while the missionary labour colony provided the scope to lead a comfortable sedentary life, which they started, from the beginning of twentieth century.

The Shenbagathope people are more individualistic, lead egalitarian life and try to solve their conflicting situation by migrating to interior isolated forest areas and thus could not adopt the disciplined life which are usually associated with sedentary life. But in Palamalai, being labourer in the estates they developed the habit or regular disciplined life. Thus they could adopt group housing projects more easily and could aspire for more such housing colonies in future.

The experience in Shenbagathope when first phase houses slowly ruined and collapsed, villagers of Shentbagathope could do nothing to repair or renovate such houses. Nor they have learnt how to build their own modern houses and lead a better life.

On the contrary, after retirement, some of the Palamalai estate labourers have encroached or got permission from Government to cultivate in 'Porombok' land and thus lead a more secured life as other caste people. In this regard 'fathers' of the mission advise their community members, how to own such Government land and lead a peaceful secured selfsufficient life.

In brief, in the present transitional phase "Group Housing" colonies are offering new opportunities to adopt a sedentary life. With the provision of more funds for construction of housing colonies and careful planning for their economic development the Paliyans can transform their present socio-economic life to a more peaceful and fruitful life style.

## Health and Family Welfare

Living at the fringe of forest and agricultural Zone, Paliyans have evolved their own indigenous medical system, which is embedded in their culture. Through age old traditions they have learnt to fight diseases and injuries in their own ways. The most extensive account on the health system of the Paliyans comes from Gardener (1997). Paliyans believe that the diseases are caused due to attacks by spirits or punishment of gods or through sorcery. The other causes are accidents, infection by microorganisms and also stress and old age

Gardener has recorded the concept, causes, remedies of various illness as existed among Paliyans. Before discussing the present development in health care system as introduced by government and Non-Governmental organization, a brief discussion of traditional medical practices will be presented below as gathered from literature and our personal observation.

Paliyans possessed their own systematic knowledge about the reasons of illness for various diseases, its symptoms and the diagnostic characteristics along with method of treatment through herbal medicine or through the practice of 'shamanism'. We have given detailed descriptions on the herbs used by Paliyans in chapter –II while discussing on medical plants.

Due to diminishing forest cover, the life pattern of the Paliyans changed and along with it the system of health care. For various reasons now they have come closer to the outside world and broke their own secluded community life and intermixed with plains people, influenced by missionaries and other outsiders.

It is true that now Paliyans are in two worlds concerning their health care. For certain situations they try to adopt their age old treatment but on certain occasions they are forced to adopt the modern medicine available around them. The traditional knowledge continued to fade from their memory. Lack of confidence, experience and attraction by Government and missionary hospital and Government schemes brought them to the allopathic system of medicines at which they are now at present.

In a study conducted recently on Paliyans of Kodai Kanal the following diseases are found to be commonly prevalent.

| Name of Disease | % of patients remote areas | % of patients in easily accessible area |
|---|---|---|
| S.T.D.<br>Tuberculosis | 24% of the respondent<br>47% | 50% of respondents<br>14% |
| Diarrhoea affecte by | | 46% of the tribal population are Diarrhoea twice a month |
| Typhoid & influenza | | 62%-65% are attacked by these disease twice a year. |

*Source:* Rajkumar R &C. Bala Sundaran (1995)

The study appears to be slightly away from the reality since it includes Pulians as well. The report also reflects on the high occurrence of S.T.D. owing to the fact that Paliyans being illiterate in general neglect the symptoms in the initial stage and seek medical advice at a very acute condition. The occurrence of high incidence of T.B. arise due to the poor and dark living conditions.

The study does not account common ailments like, cold, fever, minor stomach disorders etc.

Paliyans in and around Shenbagathope retain some of the characteristics of traditional health care till today. In a typical Paliyan hamlet even a child can identity the plants nearby with its medicinal properties. However, the income and awareness of the members of the community influence the choice of treatment.

Paliyans with relatively better economic status prefer approaching a doctor at hospital in nearby Srivilliputhur town or at some private clinics which provides them flexibility in timing and access. They also believe in better results in private clinics than Government run hospitals.

Relatively poorer Paliyans resort to traditional health care at the initial stages of the disease. If the diseases do not subside, they choose to approach a doctor as second step. We have described most of their herbal care in Chapter - II. The depth of traditional knowledge is also acclaimed by neighboring villagers who sometimes approach knowledgeable Paliyans to solve their health related problems.

Government attention seemed to elude the Paliyans till date and the Paliyans in some areas are weded to their age old traditional practices. But with the expanding welfare activities the younger generations are trying to avail the treatment and other opportunities of modern medicine. In one instance, a mobile van was provided by government for visiting the Paliyan hamlet at Shenbagathope bi-weekly. But it has been discontinued (Chapter – VI).

Case :- In one incident Paliyans affected by Cholera, rushed the patients to Srivilliputhur hospital. Neglect by hospital authorities consumed few lives. Paliyan youths reacted vehemently, which alerted the higher authorities to take preventive measures to control the epidemic.

The Paliyan hamlets at Palamalai present a contrasting picture regarding health care compared to Shenbagathope. They have the facility of approaching NGO run hospital at Kodai Kanal or the hospital run by missionaries in the plains area. Plantation estates seem to arrange the visit of a doctor to the estates at regular intervals which the Paliyans avail. Paliyans working at missionary plantations have the facility of reimbursing the cost of medicines upto 50% of expenses which is at the discretion of the missionary 'father'. Unlike Shenbagathope, Paliyans of Palamalai do not depend upon traditional medicines. It appears that they have been influenced by missionaries to seek the help of modern medicines. However, due to distance of the hospitals, Paliyans here spend a substantial amount on transportation, and other expenses.

## Quacks in Palani hills

It appears that inadequate health provisions provide a scope for quacks to exploit the Paliyans. During my reconnoiter tour in Palani hills away from Palamalai. I came across quacks with medicine bags in the road side village at Vahragaraparai and other hamlets. Paliyans affected by pain seek the help of quacks being ignorant of their skills and capability. Mostly the quacks give injection which gives instant relief and extract the fees (Rs. 5/-) Paliyans are happy at the result and do not seem to bother for long term effect.

The above study of Paliyan health system is not a specific concentrated and elaborate one but it is intended to provide visible observations regarding present health care of Paliyans in developmental context.

At first Paliyans community as a whole apprears to be passing through a transitional phase concerning their health care. They are moving away from their traditional health practices and exhibit less confidence in the old system. While Paliyans at Senbagathope still retain some of their traditional characteristics, at Palamalai Paliyans give very less priority and give preference to the modern medicine. But beyond Palamalai, Paliyans in Palani hills practice shamanism a remnant of the old system and are victimized by the quacks because of their eagerness to get relief from pain.

Secondly application and use of traditional herbs is declining rapidly under the influence of modern medicines, interaction with outsiders and increase in economic activity.

Government endeavour to the cause of health care is inadequate and fails practically to meet the demands of Paliyans. Recently the contribution of missionaries and voluntary organizations go a long way to introduce modern health care systems. There is little or negligible attention by the community towards women's and childrens health care. There is no report available specifically on this. Illiteracy contributes to the neglect of diseases causing quicker avenues to spread the contagious diseases like STD, TB etc.

## Concluding observation

The above discussion regarding the Shenbagathope area and Palamalai area and their surrounding neighbouring Paliyan areas exhibit two contrasting phenomena. Basically Shenbagathope area belongs to traditional society where modern changes are infiltrating slowly. While Palamalai represents the centre of modern changes, its surrounding area is still intertwined with past orthodoxy and traditional superstitions.

## Educational Development

It is undisputed fact that for the socio-economic and cultural development of any traditional tribal society the first and foremost priority goes to its educational development. This assumes much greater importance when the target group belongs to the seminomadic, foraging community such as Paliyans. The Paliyans render two very significant economic services. They collect immensely valuable herbal wealth from the forest and various types of minor forest produce, which are always in demand, yet they remained poor till date due to excessive exploitation by the forest contractors, business men and their employers. For their emancipation from such exploitation again, education is a must that would bring their economic development. Now they are passing through a transitional phase passing from an age old traditional life to rapidly changing era of modernization. For facing such demands of modern life they need to be equipped at least with literacy.

In this context, efforts for educational development in both the villages is presented below.

The Paliyans in Shenbagathope village even after independence remained a preliterate community. In the year 1962 the first Ashram School was established which in course of time became an Upper Primary School with five classes. But for teaching in all the five classes only one teacher is employed till now.

As regards the school, Indian express wrote " Apart from free food for three times a day, books, slates, note books writing materials and two sets of uniform are also provided to the children. The teacher has been given separate living quarters and Government sanctioned an incentive allowance of Rs. 100/ - per month" (Indian Express 3rd Jan '1984)".

The school is run by the Adi-Dravidar Welfare Department of the State Govt.

The Hindu wrote about the school "but-though there exists a school with a strength of 22 students and a hostel for the benefit of tribal children... it is alleged that the teacher is rarely seen. Whenever there is a compliant, all that the Government does is just to change the teacher. Several representation have been sent to the Government and only following the latest one to the Chief Ministers special cell, that the matter is being looked into" *(Hindu: Friday, June-20, 1987, page- 17).*

During my field work the teacher was frequently absent from the school. The day he attends to his duty he is busy with his official work or spends time in checking the mid-day meal accounts.

Most of the tribal students do not like to go to school as usually they like to be with their parents. When they are grown up to some extent, they try to assist their parents in the domestic work or do some work for the family either grazing the goats or collecting faggots etc.

One of the special programmes in Tamil Nadu to attract students is to provide noon meal in the schools. This is a very popular scheme and most of these children get opportunity of availing the noon meal scheme which is a pivotal step towards bringing the poor and needy children to attend school.

Further, the author observed in the field work that only during the time of distribution of food the children used to visit the school. On enquiry as to why no regular classes were conducted the teacher who most often spent his time idling away replied to my query, thus – "To make the Paliyan children literate and educated it is vital that they have to be systematic. So they are taught to come in time for their food and fall in line for collecting it."

Such vague and irresponsible statement of the teacher needs no comment. With the introduction of noon meal scheme attendance in school has increased substantially. But there is no qualitative improvement. The students get promotion automatically to the higher classes. Due to such liberal attitude, by the State Government students do not study sincerely. The following table shows the declining educational activities in the school of Shenbagathope.

**Table No. 5.1 – The Strength of the students in different years.**

| | Class – I | | | Class – II | | | Class – III | | | Class – IV | | | Class – V | | | Total |
|---|---|---|---|---|---|---|---|---|---|---|---|---|---|---|---|---|
| Years | B | G | T | B | G | T | B | G | T | B | G | T | B | G | T | |
| Sept' 1983 | Details not available | | | | | | | | | | | | | | | 13 |
| Sept' 1984 | Details not available | | | | | | | | | | | | | | | 24 |
| Dec' 1986 | 3 | 5 | 8 | 1 | 1 | 2 | 3 | 2 | 5 | 2 | 0 | 2 | 2 | 2 | 4 | 21 |
| 1987 | Details not available | | | | | | | | | | | | | | | 22 |
| 1993 | - | - | - | - | 4 | 4 | 3 | 1 | 4 | 3 | 2 | 5 | 2 | 2 | 4 | 17 |

**B – Boys, G – Girls, T – Total**

The above table shows how the educational level is always fluctuating in Shenbagathope. In spite of free food, dress and other reading materials the school does not educate students successfully. Some students after reading for four years, it is reported in news paper, do not know how to write their names. The school has five classes with one teacher. This might be an obstacle to teach so many classes at a time. Yet there are a few students whom even a single teacher can successfully train. But for such success the teacher should have sincerity and diligence which is found to be absent.

It is unfortunate, most of the tribal schools in interior areas, have such miserable situation and Govt. should devise some means to alter the situation.

## Palmalai School

The Palmalai School is probably the first school in whole of Tamilnadu for the education of Paliyan students. It started in the year 1915 but got its sanction in the year 1930. It is built adjacent to the church and is always under supervision of the 'father' and Missionary authorities. After passing from this school the interested students are encouraged to get admission in the missionary managed school at Shenbaganur. The students get the scope to learn discip-line and develop reading habit with proper guidance and the supervision from missionary staff. Such an old school with its long traditions brings revolutionary change in the traditions and habits of Paliyan culture. So the students of the school have progressed a lot. With their knowledge, they help in bringing social change and in integrating isolated Paliyans in to the main stream.

## DISCUSSION

The school in Palamalai is one of the oldest mission run U.P. School in Tamilnadu. It is surprising that in Shenbagathopu an Ashram school was started as late as 1962 while in Palamalai it was started as in 1915 nearly half century earlier. But it seems in Palamalai students donot get any extra benefit inspite of their early schooling for five years. As regards utility of their education for long five years, the senior members of the community frankly admitted that after completing their education, they join in the labour force of the estate and their life style changed drastically. They informed that they lost the habit of reading and writing after the heavy physical work in the estate. Thus in Palamalai the students do not derive any extra benefit through education excepting a few who opt for higher education outside their village.

In Shenbagathopu the students do not join in any secondary school after completion of their education in a school. But in recent years even with the primary school education, in later stage they are in a position to fight for their rights. We have mentioned how during prevalence of cholera when few Paliyan died in Srivilliputur hospital due to indifference of hospital authorities in taking prompt action in the treatment of such patients, the educated youths of the village strongly revolted and authorities became cautious and rendered due services towards the Paliyan patients. Further the educated youths do not allow openly non-Paliyans to collect the M.F.P. from the forest as it falls under sanctuary area. In short with education and exposure tribals are now in a position to fight against injustice and exploitation. Also they get better information and are in a position to avail new opportunities through their jobs in voluntary and Government organizations.

# 6

# Developmental Process in A Tribal Village – A Case Study

## SECTION – I

In post-independence era, the problems of backward classes, in particular the Scheduled Castes and Scheduled Tribes were considered most vulnerable and various efforts were undertaken to bring them at par with other advanced communities. Their problems were considered more serious and special efforts were taken to raise their standard of living through socio-economic and educational development as per constitutional provisions. Besides, special privileges were also granted to these people from time to time. In the previous chapter overall picture of the tribal development policies and strategies for their implementation have been discussed in general, and the nature of development in two villages viz; in Palamalai and Shenbagathope have been discussed. In 1983 a special effort was taken by the local MLA regarding bonded labour after which the tribal problems of Shenbagathope were brought to the attentions of Supreme Court. Under the direction of Supreme Court, Government of Tamil Nadu made an enquiry as regards to the existence of bonded labour and to evaluate their socio-economic condition. Though the existence of bonded labour could not be established conclusively, the poor socio-economic condition of the Paliyans, their educational and social backwardness could be focused by such enquiry. The media also played a major role in focusing attention of the Government and public on the miserable situation of the exploited Paliyans in Shenbagathope village. Two national level out-standing English medium newspapers viz; The Indian Express, The Hindu, a few Tamil Newspapers and magazines brought the matter to the forefront. Besides the media, some other agencies also played major role in this incident.

1. The major role was played by a social action group (N.G.O) which initiated the developmental process.
2. The Supreme Court of India also gave direction to the state government to discharge their duties for betterment of the exploited tribal community, i.e Paliyans.
3. The enthusiastic government officials of Tamil Nadu played a major role in discharging their duties for the development of Shebagathopu, whose development became the prior objective of Tamil Nadu government. The micro-study in a small hamlet is likely to throw light on the problems of a tribe in the village itself. Also it will be helpful in identifying and adopting better developmental strategies for tribal development in other tribal areas of India.

The presentation of this case study would be organized at three parts. In the first part, the circumstances the situation in which the initiation of the developmental activities originated and subsequent incidents of, developmental process would be described. In Part-II the governmental efforts to eradicate the problem of the Paliyans would be described. In part-III through the indepth analysis, the results of such activities will be highlighted along with the delineation of the present position of developmental process. Further the direction of the present process of change will be focussed. An effort has been made to identify the factors which accelerated the developmental process or which inhibited such efforts. Such factors would be identified with a view to achieve greater success in tribal developmental programmes in future.

## EFFORTS TO ERADICATE BONDED LABOUR

In India, in almost all states, some form of bonded labour system was existing traditionally as a part of agrarian economy. Though its prevalence was found both in tribal societies and societies of caste people, most often, caste people or those who were more acquainted with monetized economy could exploit the tribal people by advancing loan to act as labourer until they redeemed the loan. But in actual situation, once they became indebted, their loan never ended. They were so scantily paid in those days that their loan further increased but rarely ended. Thus once indebted, they remained indebted through generations. Therefore even during colonial period, efforts were made by government & social workers to help such disadvantaged innocent tribal labourers who were burdened with loan till the end of their life. But with independence of India, the country as a welfare state tried to redeem the indebted tribal labourers by framing special laws for their emancipation. In this effort, the central government directs the state government from time to time to take different measures expediously for the emancipation of bonded tribal labourers.

In one such meeting for the emancipation of the tribal bonded labourers. Thiru Thamaraikkani, the AIADMK M.L.A. from Srivilliputhur constituency made a statement regarding the widespread incidence of bonded labour in Shenbagathope village in Srivilliputhur Taluk and is only at a distance of 10 K.M. from taluk headquarters.

After receiving this information the 'Indian Express' published the news item rather in a colourful manner under the caption "Bonded Labour near Srivilliputhur, (Express News Service, Madras Oct: 1983). The extract of this news item is given below describing economic life of Paliyans.

They work in the groves and private forests around Srivilliputhur in Ramanatha puram district. Their wage is half a *padi* (less than a kg) of rice per day per family and their perks are a few betel leaves and a little tobacco or snuff.

There are about 200 of them, all Palliar tribals, who have, for ages, been toiling as bonded labour.

About 50 of them are working on the lands in Shenbagathope,....... Which also includes about 250 acres belonging to the Srivilliputhur Andal temple but currently leased to a private party.

Though bonded labour has been in existence in this area for years, governmental agencies seem to have done nothing about it.

To alleviate the problems of the tribals and to improve their standard of living tribal Co-operatives had been set up here but the benefit derived from these Co-operatives donot reach the tribals as the landlords and other vested interests have filled these Cooperatives with "Benami" members, it is alleged.

A few years ago the government constructed a few house for these tribals at Shenbagathope, but these houses stand in ruins, save for a couple of them. The roofs have collapsed, the walls have rumbled.

Officials claim that the tribals did not move into the houses at all. The Government Tribal Residential School (G.T.R.S) at Shenbagathope is a stark testimony of how well intentioned scheme can be brought into a naught by the mismanagement of employees and the in-difference of higher authorities. Take the case of ten year old Veeramma, she is supposed to be studying in IV standard, but she can not even write her own name.

X X X X X X

> During the past five years at least seven children have died due to unknown causes. Local A.I.A.D.M.K. M.L.A R. Thamaralikkani claims these deaths have been primarily due to malnutrition.
>
> And while the exploitation goes on, the Poliyar tribe which is confined to Srivilliputher range is on the brink of extinction as the mortality rate is high among them. *(The Indian Express: 1983).*

The above news item in Indian Express created some amount of alertness in a voluntary organization which took immediate action. "On the basis of this news item a telegram was sent to the Supreme Court of India by Managing Trustee, Society for Community Organization Trust (SOCO), Madurai with a request to treat this telegram as writ petition which was followed by a detailed letter to Honourable Justice P.N. Bhagawati (Judge, Supreme Court of India). Supreme Court of India treated this as writ petition and the court in its record of proceedings dated 12th Dec. 1983, after stressing the importance and necessity of associating social action groups in conducting inquiries into matters of this kind," suggested to include a representative of the petitioner. As per the direction of the Supreme Court an enquiry committee was formed with the following members.

1. Collector, Ramanatha Puram District
2. D.I.G. of Police
3. Managing Trustee of the voluntary action group (SOCO).

## Report of the Enquiry Committee

After formation of the team, the members tried to immediately pursue forward their objectives without loss of any time. Supreme Court ordered for the formation of the inquiry team on 12th December 1983 and in less than a month's time the team was ready for investigation. By that time SOCO had already posted an executive of their organization who had already investigated the issue established rapport with Paliyan labourers of the village, visited the surrounding areas and had collected information about Paliyan labourers.

On a Sunday morning i.e. 8th January 1984 the team started for Shenbagathope with the officials of several departments along with armed police. The villagers of Shenbagathope had already gone to their fields to carry out their usual work in a harvest season. The arrival of police force stunned the labourers (as reported by an observer). However they explained the purpose of their visit and all of them were brought to the village Shenbagathope for conducting their enquiry. The team interviewed about ten Paliyan families as per the enquiry report.

## Impression of the team as regards Paliyan labourers

As reflected in the report, the team could realize the miserable condition of the agricultural labourers & their ignorance.

The very appearance of these people showed that they were not only extremely poor but were also totally in the dark about the outside world. Almost all of them are working as labourers in either *topes* and lands belonging to Arulmigu Nachiar Devasthanam of the temple in Srivilliputhur in which mango, coconut, tamarind and fruit bearing trees are leased out once in three years or similar private *topes*. All the labourers are illiterate and are quite ignorant of their rights and entitlements as agricultural labourers. *(Enquiry Report: 1983).*

## Lack of Indebtedness among the tribals of Shenbagathope

The bonded labourers usually lead a bonded life due to their indebtedness to their employers. But in case of Paliyans of Shenbagathope all Paliyans denied that they had not taken any loans from their employers. The enquiry team mentioned out that their living conditions were comparable to those of bonded labourers. Thus they wrote :

Even though these villagers are not indebted monetarily to their employers as in the case of usual bonded labourers still the conditions of their living are comparable to those of the bonded labourers like working for nominal wages or lack of ability to other parts or any viable alternatives in the place of residence to eke out their living.

## Poor living conditions of the Paliyans of Shenbagathope and nature of their wage

The team observed from the interview with the labourers and their families, that they were being paid wages only in kind at the rate of one measure of broken rice (1.75 Kg) per day of work. In addition, they are also given some tobacco betel leaves and snuff. They seem to work only for a period of three months in a year. They are not being paid any wages in cash at all. They are given one set of Dhoti and Towel during the course of a year. During the rest of the year, i.e. for nine months' period, these people are living merely on some root vegetables. The market value of the wages in kind paid to them would not be more than Rs. 5/- for men and Rs. 2/- for women. These families have been in this kind of pre-occupation for the past several generations. (First Enquiry Report :1983).

Further the enquiry states "In fact most of them do not seem to have handled cash and several of them were not able to distinguish a two rupee note from a 10 rupee note".

Further the enquiry revealed that though they were not monetarily indebted to the employers still they had suffered ill treatment at the hands of their employers. Stray instances of the labourers of having been beaten by their employers for not turning upto work in time or their having to go to the hospital were also reported and the police officials were asked to register such cases.

The team also found that most of the labourers work on lease lands and the lessee have been changing once in three years. Still one has to recognize the fact that the lessee who have been successful in obtaining the lease from the temple authorities in the recent past are related to each other closely. Again, it is mostly a single family who have been keeping effective control over these illiterates for the last six years. Thiru Durairaj Nadar of Mamsapuram was the lessee from 1-7-99 to 30-6-1980 and has again become the lessee from 1-11-1983 to 31-10-1986. Thiru P. Seeniappa Nadar was the lessee from 1-11-1980 to 30-11-1983 who also belonged to the same family *(First Enquiry* Report: 1983).

The team drew the attention that there were no medical facilities in the village. The nearest medical facility in Srivilliputhur was at a distance of 10 km. away from the village. They further mentioned that the Paliyans could not avail the medical facilities at Srivilliputhur as people did not possess any money towards bus charges. They were of the view that all the members of the family looked underfed and few of them looked healthy.

After collecting data the team gave the opinion that Government was aware of the problems of Paliyans in Shebbagathopu and had taken steps to alleviate their sufferings. They had constructed good black topped roads to Srivilliputhur few years ago. A Co-operative society was formed but it collapsed as outsider entered into it. The team suggested that the same Co-operative should be revived taking only people of hill tribes as its members. The role of the school was discussed while exploring how the children in the school are given clothes, books & food by Harijan Welfare Department. There were 13 children in the school and the health of the children looked quite good. The students were attending school regularly. The condition of the school building was also quite good. Due to interior location of the school there have been frequent changes of teachers. Even though Government have constructed some houses for the tribals they have chosen not to occupy the quarters and preferred to live in their own huts. Thus the team believed that the Government had taken some measures for the welfare of tribes. Further some recommendations had been sent to the government as part of the working group report on tribal development during the plan period. So enquiry committee clarified that the working group report was sent on 28th May 1983 much before the publication of the condition of the Paliyan in The Indian Express in Oct'1983. Some of the recommendations were interesting.

The license for collection of honey and minor forest produce should be given to tribals. Disposal of the forest produce through Co-operative basis, training of women folk in the manufacture of essential domestic articles made from bamboo and formation of a multipurpose co-operative bank were the other recommendations.

Finally the team recommended measures to be taken on priority basis in a phased manner for the upliftment of the tribals.

## SECTION – II

## ACTION TAKEN OVER THE ENQUIRY REPORT FOR THE DEVELOPMENT OF PALIYANS OF SHENBAGATHOPE

After the submission of the enquiry report to the Supreme Court, there was discussion at different levels between Supreme Court and Government of Tamil Nadu. Eventually, Govt. of Tamil Nadu assured Supreme Court that it would take all steps for the development of tribal bonded labourers in the state. Accordingly District Magistrate of Ramanatha Puram was directed by the state government to take adequate steps for the development of Paliyan labourers as per "The Bonded Labour System (Abolition) Act, 1976 which states that :

The District Magistrate is authorized by the state government............ And the office specified by the District Magistrate shall, as far as practicable, try to promote the welfare of the freed bonded labourers by securing and protecting the economic interest of such bonded labourer so that he may not have any occasion or reason to contract any further bonded debt.

Incidentally the then district collector who was asked to take charge of the welfare of the freed bonded labourers and entrusted with protecting economic interest of such people acted enthusiastically for the development of Paliyans of Srivilliputhur Taluk.

On the 19th January 1984 there was a news item published in the Indian Express for the public. The contents of this news item find its origin from the enquiry report.

The Collector informed Mr. Thamaraikani that there was no bonded labour system enforced on Paliyans, a tribal sect living in Western Ghats abutting Raja Palayam Srivilliputhur and Watrap areas ......... 92 families of Paliyan living in various places in Western Ghat range had employment only for three months in a year and depended on natural resources like roots for their livelihood during the remaining nine months.

He said that to improve their condition a crash programme was being chalked out at the district level. Steps have also been taken to electrify the tribal school building at Shenbagathope. Opening of a noon meal center and introduction of transport and medical facilities were included.

They also examined the possibility of assigning land to these tribals and also to provide employment for them in the proposed herbal farm to be started in Shenbagathope area. Bonded labour issue was not only discussed through media, but also Mr. Gurumurthy (the Collector of the Kamarajar District) was taking urgent steps for improving the conditions of the Paliyans speedily. As District Magistrate, he had the authority over all district level officers and thus could influence them for advancing the cause of Paliyan welfare. After the publication of the news item regarding the proposed crash programme to improve the condition of Paliyans, District Magistrate held a meeting of all district level officers who were in charge of their respective areas.

Minutes of the meeting were to discuss the welfare measures for the upliftment of the tribal of Shenbagathope, Ayanar Kovil (temple) and surrounding villages in Srivilliputhur and Raja Palayam. The meeting was conducted by the collector of Ramanathapuram at Madurai on the 23rd January 1984.

1. **Medical facilities and establishment of mobile health camp:** District Medical Officer, Ramanathapuram reported that the Medical needs of the tribals of Shenbagathope and surrounding

villages would be attended as they would be visiting the villages on every Monday and Friday, with the mobile van.

2. **Bus facilities:** The District Transport Officer agreed that a town bus would make one trip in the morning and one trip in the evening. Road between Shenbagathope to Mamsapuram to be taken up for black topping by Highway Department out of the funds provided by the Director of Adi-Dravidar and Tribal Welfare, under the Tribal Sub-plan.
3. **Old age pension (OAP) cases:** The R.D.O, Sivakasi reported that, OAP has been sanctioned for 3 people of Shenbagathope.
4. **Noon meal centre:** The Noon-Meal Centre was opened in the village U.P. school (Senashram) of Shenbagathope. Besides the school children, the old people who were getting old age pension were allowed to get meals from the Noon Meal Centre.
5. **Allotment of Land:** The District Revenue Officer reported that about 293 acres of Porambakam lands in Sl.No. 1 of Muttupalli village were available and there were about 1955 trees likely to be cut which would be useful for match industries.
6. **Formation of co-operative society:** The Joint Registrar explained in detail the reasons for the failure of old society and promised to revive the same by the end of the February'1984. He assured that the society would be composed of tribal people only and the Forest Ranger would be the president.
7. **Khadi & village industries board:** As the tribals are well versed in honey making the Asst. Director, Khadi, Srivilliputhur stated that Khadi Board could supply bee hives to them & train them in bee keeping. The total cost was reported to be Rs. 41,200/- and the Project Director RDDA promised to meet the cost of training i.e. Rs. 65,000/- out of TRICEM funds. The training was to start from the 1st February 1984. The training of women folk in basket making from bamboo could also be done under TRICEM and the funds could be arranged by the Project Director RDDA.
8. **Milch animals:** As the response of the tribals to rear milch animals was positive, it was decided that Deputy Registrar (Dairying) should identify the tribal beneficiaries and send proposals for supply of milch animal in that current year itself. It was decided to recommend to government, as a special case, for relaxation on the ban of giving buffaloes after their first delivery in the special circumstances of this case. The Managing Director

P.D.M.A.P.C. was entrusted to see that the milk is purchased by the Srivilliputhur Milk Society directly from the tribals by arranging to collect the milk from the village. The cooperative peon was assigned to milk the buffaloes and to carry the milk to the co-operative store at Srivilliputur.

9. **Horticulture:** Joint Director, Horticulture promised to provide employment for Paliyans in Poovani Orchards.
10. **Issue of family cards:** It was decided that as a special case, each tribal family would be provided a family card for the supply of rice etc. The proposal was sent to the Commissioner of Civil Supplies.
11. **Improvement of the working condition of the school:** It was decided to improve the teaching in the school by providing quarters for the teacher and by electrifying the school.
12. **Repairing of houses:** The houses constructed by the Director of Adi-Dravider in the year 1965 were to be repaired.

Besides such meetings consisting of District level Officers to discuss the crash programmes and to arrange the funds for different items, it was suggested to make mass contact programmes in Shenbagathope and its surrounding villages.

Thus while the District Magistrate was busy in organizing activities for the welfare of the Paliyans, SOCO was also busy in its own way for the welfare of the Paliyans by disseminating information relating to Paliyan welfare activities through the newspapers etc.

After sending the telegram to the Supreme Court with a request to treat it as a writ petition, SOCO sent a worker to Shenbagathope to identify the bonded labourers and to study their problems and to gather information regarding the local situation. During the enquiry, the labourers insisted that they were paid in cash towards their wages. This was denied by a student of Class-V. It can be presumed that the labourers trained by their employers to emphasize on cash payment were cross checked with denomination of Rs. 10 and Rs. 2/- in hand to which the labourers failed to recognize the notes.

After drawing the attention of Government to the miserable condition of the Paliyans, SOCO tried to improve social status of the Paliyans by modifying their social customs. In this regard the Managing Trustee tried to introduce arranged marriage and provided saree & other clothes for the marriage ceremony. But he admitted that he was not very successful in reforming the Paliyan marriage customs. SOCO also tried to bring scattered Paliyans from different villages and concentrated them in Shenbagathope so that developmental efforts may yield them greater amount of benefits.

## SECTION - III

In the previous sections it is observed how the developmental process was initiated and implemented through SOCO and Government and how through the directive of Supreme Court Government undertook sincere efforts to investigate the conditions of the Paliyans and could plan for their emancipation from poverty and exploitation. Yet behind the apparent success stories there are incidences of failure of such efforts. In this part the success and failure of developmental effort will be analysed with a view to identifying the loopholes that could be avoided and to focus the precautions that could be adopted with periodic evaluation from time to time. Finally the paradoxical situations that negate the success would also be explored.

In the second enquiry report prepared jointly by Collector, Ramanathapuram and Managing Trustee of SOCO, the major developmental activities pursued by the Government have been elaborately presented. Here description of developmental work will be presented in brief followed by analytical discussion over these activities pursued by Government. The author would try to organize the actions of Government on economy, education, health and other welfare activities. Since economic backwardness is the major issue of tribal society particularly in the seminomadic Paliyan Community, discussion on economic development was taken up at the first instance.

## ECONOMIC DEVELOPMENT PROGRAMMES

**Honey extraction:** Honey Collection is one of the major traditional occupation of the Paliyans. So the collector of Ramanathapuram while planning for income generating projects gave emphasis on honey extraction. The following para gives the details of honey extraction work in Shenbagathope.

"The training in bee keeping industry for 30 people under Integrated Rural Development Programme for a period of three months was imparted at an expense of Rs. 22,500/- towards stipend, cost of material etc. After completion of the training each beneficiary was distributed with two bee-hives and they have been installed at various places." *(Second Enquiry Report, 1984).*

It was said that honey would form in the hives within three months and the flowering season was reported to be January to June of the year.

**Comment:** Though honey collection is a traditional occupation of Paliyans, their process is totally different from honey production through bee keeping and honey extracting from the bee-hive boxed. It requires patience, efforts for the observation of the behavior of the bees, which are placed in boxes. Bee hives need inspection everyday. Excess queen eggs are to be destroyed or else when matured they may drive out the old Queen.

The trainees naturally had to wait for three months during which no honey could be collected. There was no necessity for training a large number of trainees. Most of them hailed from several villages and were brought to Bhagavatinager (renamed village of Shenbagathope). When they could not find any regular source of income they returned to their village. As the Paliyans could not adopt to settled lifestyle in their new houses, it was difficult for them to have a daily or periodic observation of the hives. During the field work the author could not meet even a single person who had adopted honey collection even as an auxiliary or part time occupation. At present no one knows traditional honey collection. Those who were trained in bee rearing forgot every thing that they learnt during training period.

**Training in basket making for women folk:** Twenty people were trained in basket making for a period of six months. They were paid stipend of Rs. 125/- p.m. each and Rs. 4000/- had been provided for the purchase of raw materials and tools at the rate of Rs. 200/- each. They became members in the Multipurpose Co-operative Society, Srivilliputhur. The Multipurpose Cooperative society agreed to undertake the supply of raw materials and the purchase of the finished products. The women folk who were engaged in basket making replied to the effect that they had just started the work and that they earned Rs. 2/- per day.

**Comment:** It is not known to what extent these women could manufacture baskets. But villagers say that after manufacture of these baskets it could not be marketed and hence women discontinued the manufacture of baskets. According to the enquiry report, Multi-purpose Co-operative Society of Srivilliputhur was to arrange the supply of raw materials and purchase of the finished products. Why the above named Society could not discharge its obligation was not clearly known. It seems no body took the responsibility seriously, leading to failure of the project.

**Cultivation of medicinal herbs:** The Director of Indian Medicines had made a programme to cultivate some herbs in this area. For this purpose, they identified an area of about 33 acres and took the land on lease from Arulmigu Andal temple. The Paliyar tribals numbering about 25 persons were to be engaged daily and they would be paid wages at the prescribed rate.

TAMPCOL requested the District Collector for a house to store the implements and stock of herbs and for the accommodation of officials. The District collector agreed and authorized one of the rooms in the teachers' quarters for immediate use with the promise to allot a house soon after.

**Comments:** TAMPCOL collects a large quantities of medicinal herbs through the Paliyans of Shenbagathope which are transported to Tirunelvelli for manufacture of Ayurvedic medicines. The wages that they get from TAMPCOL depends on quantity and quality of such herbs. The man who gets highest amount of wage for such collection of herbal medicine before the field work was getting Rs. 50/- per day. But this man usually takes frequent leave. No visible consequence of medicinal herb cultivation was observed.

## Land Allotment

Since the Paliyans are a food gathering seminomadic community, the permanent solution for eradication of their poverty lies in providing agricultural land for their permanent settlement. They are not willing to move out of their existing habitation and settle on the plains. It was therefore desirable that any land assignment to the Paliyans should be in the vicinity of their habitation. Keeping in view the above fact, forest Poromboke lands measuring 61.56 acres have been selected for assignment. At present, the lands are not fit for cultivation and there is no water facility. To reclaim the land so as to make them fit for cultivation the District Collector has sent proposals to Government at an estimated cost of Rs. 4.75 lakh. Orders of Government are awaited. Further, discussing about the leasing lands belonging to the Arulmigu Andal temple, the enquiry team suggests, " It is considered that getting of lease of some private hold lands will not be a permanent measure to uplift the economic condition of the tribals. On the other hand, if the government land is assigned, they would be encouraged to cultivate the lands on a co-operative basis and there would

> be scope for employment throughout the year *(Second enquiry report, 1984).*

Comments: The importance of land for the Paliyans had been discussed several times. In a meeting, organized by the collector of Ramanathpuram at Madurai on the 23rd January' 1984 (as per minutes of the meeting) District Revenue Officer reported that about 293 acres of Poramboke land in Sl.No. 1 of Mettupalli village was available and there were about 1955 trees likely to be cut which would be useful for match industries. In this way there had been discussion to give land to Paliyans but so far no step had been taken to hand over such land to the Paliyans for the permanent solution of their poverty.

The Hindu (June 26th 1987 issue) writes, "Topping all the demands is the strong plea for the immediate provision of about two acres of land to each tribal so that he could cultivate on his own. According to Muthu Irulandy, there are enough lands at Kulavanparai located nearby and the government can think of allocating the land to them. In this connection the SOCO spokesman pointed out that about two acres of land had been allotted to each of the labourers in Madhya Pradesh who were once reportedly bonded at Keezhayur in Madurai district.

This indicates that both government and voluntary organization, social activists not only realize but actively highlight the problems and offer proposals to eradicate the poverty of the paliyans through agriculture yet it is surprising why the government cannot provide the land when these are available near to their habitation. It is really difficult to identify the sociopolitical forces which hinder the development of poor Paliyans who are now undergoing the transitional phase from food gatherers to the stage of peasantry.

## Formation of Co-operative Society

There was a Co-operative society named "Shenbagathope Hill Tribals Labour Contract Co-operative Society". It started in 1965 but became defunct due to heavy loss. When the team visited the Shenbagathope on 8.1.1984, it was considered at that time that instead of forming a new co-operative society, the existing society could be revived and reactivated. Accordingly, several measures were taken to revive the society. Owing to the continued efforts of the district collector, the defunct co-operative society started functioning with an executive committee consisting of 9 members. Among the nine members, 5 members are tribals and the remaining four are the Ex-Officio members by virtue of their official capacities. The Forest Ranger and the Panchayat Union Commissioner, Srivilliputtur have been nominated as president and vice-president respectively and Executive Officer, Arulmigu Andal Thirukoil Devasthanam, Srivilliputhur and the Extension Officer for Co-opertion, Srivilliputhur are the Executive Committee members. Full time employment of a Co-operative Sub-Registrar as Secretary to the society had been recommended to the Govt. The area of operation of this society was confined to Srivilliputhur forest range in Srivilliputhur Taluk, which has been extended to Rajapalyam.

The objective of the society was to promote the economic interests of the hill tribals, by obtaining forest lease for the collection of minor forest produce and providing regular employment. The Government had given a sum of Rs. 10,000/- as subsidy towards the capital outlay and a sum of Rs. 2,500/- as subsidy for the purchase of bullock carts and the other implements. A sum of Rs. 4000/- had also been provided by the Ramnathapuram District Co-operative Central Bank for the year 1984 towards cash credit loan. During the year 1984, 21 tribals have been newly enrolled as members.

A license for collection of minor forest produce from Srivilliputhur forest range unit has since been granted to the society.

**Comment:** Of all the projects meant for alleviation of poverty of the Paliyans the formation of tribal co-operative society had the highest capability for improving the economic well being of the Paliyans. The Dairy development programme and procuring land for agriculture no doubt could occupy second and third priority respectively. Collection of minor forest produce is the age-old occupation of the Paliyans. Through generations they have acquired this skill and have acquired monopoly in this profession. Further it requires no capital investment. With their experience in collecting, transporting and storing they could prosper. Since they had no rival for collection it was an advantage. They faced the difficulty while selling their goods and are victimized and exploited by unscrupulous traders. The cooperative was the single best method to fight against such exploiters. Further with such an organization they could draw the attention of Government and could enhance their economic security. But we could see that the only difficulty lies in successful organization of this institution. Previously we have seen that though it started in 1965 it became defunct and Registrar Cooperative Societies explained that plains' people became members in that organization and through "Benami" transaction misappropriated the capital. Therefore, he suggested that the revived institution should be constituted only with tribal Paliyans. But it is observed that the committee was dominated by non-Paliyans. The executive committee consist of nine members of which five were tribals and the remaining four were ex-officio members by virtue of their official capacities. Thus the real power was in the hands of non-tribal members who had all the power to plan, sanction and formulate the policies. But it is not the question of tribal and non-tribal members but who are honest or dishonest members in the executive committee. The author had no access either to scrutinize the details of accounts or to enquire the members of this co-operative society. But it appeared that the initial financial support of the Government is quite encouraging. As reported by a SOCO worker, who was, directly or indirectly, involved with the day-to-day activities of this cooperative society, they made a peculiar rule for the collection of MFP. Since all the workers were members of the co-operative society they were asked to accept Rs. 6/- only as their subsistence allowance and were assured that their dividends would be distributed later. From the enquiry it is learnt that even within the first two years no dividends were distributed as there was loss in the second year due to bad weather condition. Thus there was no profit to be distributed. Enquiry could not be made into the records and account books, since most of these were maintained in local language. Besides the names of the workers who were collecting these MFPs could not be found. It was also not know to whom they were selling and at what rate. The collections were made and stored in the '*Medu*' and were transported to the purchasers. But no details of such operations were available.

After certain years Goverment restricted collection of minor forest produce as the Srivilliputhur forest range was reserved specially for flying squirrel, a rare and endangered species. With this restriction, forest department prohibited collection of minor forest produce formally. So the auction of forest beats was automatically stopped.

It appears, in the beginning of revived Co-operative society, forest department was directly giving the collection rights to Shenbagathope Co-operative society to collect minor forest products but later the forest department did not give this privilege to the same. In the Hindu (June 26, 1987) the following extracts indicate the situation after the revival of the defunct co-operative society.

> "Under one of the schemes of rehabilitation of the tribals the defunct Shenbagathope Hill Tribals labour contract co-operative Society had been revived and its

> area of operation extended from the Srivilliputhur forest range to the Rajapalayam range also. The society was to obtain lease from the forest department for collecting minor forest produce and thereby ensure regular employment to the tribals. But it seems the tribals are not generally satisfied with the society, said Mr. Muthu Irulandy, a social worker, and an SOCO spokesman. Mr. N.S. Subramanian stated that as per the rules only the tribals themselves should be the executive members of the society. But now only a few members were tribals. More over, there was also an allegation of under-payment to the tribals for the products collected from the forests after great toil."

According to Mr. Subramanian, the discontentment was very much evident from the fact that about two months ago, the tribals had requested to give them, the lease for the collection of fruits from the tamarind trees in the Alagarkoil area for Rs. 7000/- and this was accepted. But a sum of Rs. 16000/- was demanded later on. The amount was too high for the poor tribals and so the lease was not offered to them. It was alleged that an outsider had taken the lease for Rs. 16,000/- for all the beats within the Srivilliputhur forest range as against only one beat offered to the tribals for the same amount. Why should there be such a discrimination ? asks Mr. Muthu Irulandy. In fact the same opinion was also endorsed by Chinnapalani, a tribal. The offer of the lease to the tribals could have improved their economic condition, laments Chinnapalani. "There is now no proper employment for our community." All he would plead was that some effective steps should be taken to provide them with a livelihood.

**Supply of essential commodities:** Since there was no grocery store in the village, Government planned to open a fair price shop for the Paliyans. Its progress is recorded below as observed from the second enquiry report.

A fair price shop has been opened since February 1984. A mobile van carrying the essential commodities like rice, dal, kerosene, palmolein oil etc, is coming to Shenbagathope and supplying to the tribals once in a fortnight at the time of disbursement of cash towards the cost of milk supplied by them. With the income, the tribals are repurchasing them, through this fair price shop. The tribals expressed their satisfaction about the quality of commodities supplied at the fair price shop fixed by the Government (Second enquiry report).

**Comments:** The fair price shop was running for a year or so. But developmental activities nearly provided enhanced income to the labourers. As a result no one purchased from the store but preferred to go to nearby town for this. The bus was coming to the village twice a month and except subsidized articles (provided through public distribution system) people would have found it more convenient to purchase their daily needs either from Srivilliputhur or from any other convenient place. So with the closure of fair price shop there was not much inconvenience. The net result was that people's habit of depending on roots, yams, green vegetables were reduced and rice became the staple food of the Paliyans and their gleaning habit for collection of food was drastically reduced.

## Distribution of milch animals

As per the suggestion and findings of enquiry report for the development of Paliyans in Shenbagathope, twenty tribal beneficiaries were sanctioned Rs. 2500/- each on 4.7.1984 under Integrated Rural Development Programme (I.R.D.P). as the District Collector was one of the members of the enquiry team, he took special interest to accelerate the work. The Deputy Registrar, Dairy Development was asked to identify the tribal beneficiaries and to send a detailed proposal. Subsequently the Project Director (D.R.D.A) hastened the process for realization of funds. After the animals were procured, the veterinary doctor from Srivilliputhur and the medical officer of Ramanathpuram "district cooperative milk producers union" were entrusted with the health care of these milch animals. Care was taken to procure the milch animals with their first calf as it was accepted fact that the milk yield was higher in the first birth itself. Srivilliputhur Co-operative Milk supply society had the responsibility of arranging a staff for collecting the milk in a bicycle twice a day. He was instructed to milk every buffalo and measure that quantity at each beneficiary's house. All the tribal inhabitants in the newly created colony were included as the beneficiaries.

(Though there was a small hamlet here yet through the intervention of the voluntary agencies and government agencies some more Paliyans were persuaded to settle in this new colony. They were also included in the list of beneficiaries).

It was stated that the buffaloes were given in stages. Initially they were given the buffaloes in two stages i.e. 7 & 7 in an interval of two months. Subsequently six more new members were given in the second disbursement. This second disbursement also included some of the members of the first category as the concerned authorities were satisfied with their performance. As a result some people got two buffaloes and some got only one.

**Table No. 6.1 – DISTRIBUTION OF MILCH ANIMALs AMONG PALIYANS IN SHENBAGATHOPE UNDER I.R.D.P**

**Loan Amount – Rs. 2500/-**

**Thrift Capacity – 5% of cost**

**Subsidy – 50%**

**Date of sanction of First Loan – 4.7.1984**

| Name Beneficiaries | Total No. of cattle Received | Pending loan | Type of Calf received | | Milk supplied to Co-operative society in litres | | | | | | Total Milk deposited in the society in litres |
|---|---|---|---|---|---|---|---|---|---|---|---|
| | | | 1st batch | 2nd batch | 1984-85 | 1985-86 | 1986-87 | 1987-88 | 1988-89 | 1989-90 | |
| 1 Tayammal W/0 Krishnan | Two | 1 pending | Male | - | 380 | 195 | 500 | - | - | - | 1075 |
| 2 Chellaya S/o Suppan | Two | 1 pending | Male | Female | 1256 | 1286 | - | - | - | - | 2542 |
| 3 Sadeyan S/c Velayan | Two | 1 pending | Male | Male | 1030 | 325 | 1010 | 321 | 230 | - | 2916 |
| 4 Periasangeya S/o Perumal | Two | 1 pending | Male | Female | 526 | - | 506 | - | - | - | 1032 |
| 5 Pechiammal W/o Subramaniam | One | Nil | Male | - | 1140 | - | 1120 | - | - | - | 2260 |
| 6 Papa W/o Mariappan | One | Nil | Male | - | 586 | - | 967 | - | - | - | 1553 |
| 7 Ponnusamy S/o Suppan | Two | 1 pending | 205 | 185 | - | - | - | - | - | - | 390 |
| 8 Sithai W/o Seengappan | One | Nil | - | - | - | - | - | - | - | - | Dead |

| | | | | | | | | | | | |
|---|---|---|---|---|---|---|---|---|---|---|---|
| 9 Suppammal W/o Kottai | One | 1 pending | Male | - | 140 | - | - | - | - | - | 140 |
| 10 Thangavelu S/o Talamlai | Two | 1 pending | Female | Male | 774 | 595 | - | 597 | - | - | 1966 |
| 11 Chinna Palani S/o Talaimalai | Two | 1 pending | Male | - | 220 | - | 240 | - | - | - | 460 |
| 12 Papa W/o Palani | Two | 1 pending | Male | Female | 987 | 1320 | 676 | - | - | - | 2983 |
| 13 Rasammal W/o Chinnasangeya | One | 1 pending | - | Female | 394 | 185 | 415 | 185 | 166 | 236 | 1581 |
| 14 Ammeyappan W/o Shokkan | One | 1 pending | - | - | 510 | 270 | 586 | 270 | - | - | 1636 |
| 15 Karupaya S/o Velasamy | Two | 1 pending | - | - | First cow gave less milk 2nd cow died of miscarriage | | | | | | |
| 16 Ponnaiah | One | 1 Pending | 468 | - | - | - | - | - | - | - | 468 |
| 17 | | | | | | | | | | | |
| 18 Tangammal W/o Velasamy | One | 1 pending | Male | - | 283 | 87 | 304 | 185 | - | - | 859 |
| 19 C. Velan S/o Ponneah | Two | 1 pending | Male | Female | 712 | 385 | 732 | 383 | - | - | 2212 |
| 20 Muthaya | One | 1 pending | - | Female | - | Data not available as he sold his buffalo | | | | | |

Before the introduction of the milch animal scheme some of the Paliyans had the experience of grazing cattle for the inhabitants of the neighbouring villages. Initially under encouragement from the N.G.Os, Co-operative society and government officials they had a smooth sail for one year. During this period on an average a buffalo yielded 3 liters of milk every day. The Paliyans used to graze the animals in the adjoining grove and in the nearby forest areas. Apart from that in the initial stage they fed the animals with the mixed and standardized grass provided by Srivilliputur co-operative society. But this supply was for a shorter duration. All the beneficiaries were enrolled as members in the Srivilliputhru cooperative milk supply society. The society's role was to collect the milk from the tribals by milking the buffaloes & paying their dues after deducting money towards loan. The instalments included saving, cost of grass, share capital and loan amount. For collection of the milk an employee of the co-oerative society used to visit the hamlet every morning & evening. A litre of milk was taken at the rate of Rs. 3/- to Rs. 3.20.

In the year 1984-85, out of the 20 beneficiaries 16 people were found to have sold their milk to the Srivilliputhur Cooperative Society. Sithammal (W/o Singappan) who was one of the beneficiaries & was dead during field studies was one of the four who didn't contribute. Karupaya, S/o Velasamy another defaulter was unfortunate as his cattle had died of miscarriage. The third person Mr Muthaya sold off his cattle for immediate cash and the fourth person had deserted this village. Sadeyan S/o Velayan had given 1030 litre of milk to the society and he was one of the three beneficiaries who had given more than 1000 litre in a single year. Of the remaining two persons one is Pechiammal (W/o of Chellaya). Interestingly Sadeyan was the person who had got the experience of rearing cattle & sheep, much earlier to the introduction of the milch animal scheme. Unfortunately in the following year the number of participants came down to half. Probably this might have taken place due to the gestation period of the other buffaloes. In the following year 14 beneficiaries contributed more than 1000 litre milk to the cooperative milk society. This proved that these people took care of their animals. But slowly the trend to sell the buffaloes continued and therefore the number of contributors supplying milk to the society came down. In the year 1987-88 only 8 beneficiaries contributed milk to the society and in the year 88-89 it came down to only two persons.

When the village was visited in 1992 April none had any buffaloes left with them. Thus the total collapse of the project took place in five years time.

As mentioned earlier these welfare and developmental activities had attracted a lot of media attention. People in the neighbouring caste villages were more curious with sudden flow of fortune to their neighbours. Usually they exploited their labour & hence they kept them under their control. But now it was a different atmosphere. Two different voluntary agencies operated in the village. While one was concerned with the legal aid programmes, the other had taken up work on health care & development. Both the agencies had their staffs here. Moreover higher executives of government & the N.G.O.s, visited the village at times. Under this changing scenario the neighbouring villagers did not interfere directly with the Paliyans. But then they never failed to discourage the Paliyans that these activities did not suit to their traditional occupation. Meanwhile some also started selling their buffaloes as they were lured by hard cash that they received from buyers. The first seller was the boy who was working for a N.G.O. in the village. He had sold off his buffalo in the first year itself. Slowly others followed. Many attributed their sale of buffalo to the consistent complain by the nieghbouring mangogrove owners/contractors, that the buffaloes grazed in their area and they did not want this to happen. Though this might not have been the only factor in selling the animals yet factors such as discouraging the paliyans by the neighbouring caste people, allurement of hard cash were some of the strong factors in motivating them to sell off the animals. It is reported by the Paliyans that even agent of one N.G.O. started to mediate with the Paliyans for the sale of their buffaloes with the caste people. Though this man was one of the active spokesmen for the Paliyans and tried to take care of the programme for the success of it, he changed his mind for unknown motives.

The government officials had taken care to supply the Paliyans with cattles which were supplied to them with calves on their first birth so as to get a better yield. The average yield of milk in a day couldnot go beyong 3-4 litre. Though the Srivilliputhur milk co-operaive department had taken responsibility of supplying cattle feed it was only the straw that the Paliyans gave to the cattle's. Another notable failure from the Government was that they did not take care to select buffaloes with she calves. Many of the nebeficiaries got male calves and the scope to continue the practice of rearing the milch animals faced a great obstacles. Again it was reported that the visiting veterinary officials took the interest in getting the male calf sold but failed to advise the care of the calves needed. So, many calves had died off in the early stages. Eight beneficiaries had got two buffaloes in contrast to 12 beneficiaries who had received one each. These beneficiaries obtained their second loan as they had a better repayment capability. All these eight had paid back their money for the first animal. But their second installment of loan was not paid. Subsequently their money was waived off due to certain policies adopted by Government of India.

The details of the loans as regards milch buffaloes and the milk supplied to the Srivilliputhur milk co-operative society are given in the table 6.1.

## Goat distribution scheme

For economic development of the Paliyans of Shenbagathope, besides the milch animals, loans were also given through Pandayan Gramya Bank, Vilupanoor to rear goats. Each unit consisted of 21 goats or sheep at a cost of Rs. 4500/-. Out of this one third or Rupees 1500/- was given as subsidy to persons who got the sheep loan, Five people, as detailed below were granted the goat loan of Rs. 4500/- each.

**Table 6.2 – The goat loan at Shenbagathope**

| Sl.No | Name of beneficiaries | No. of units | Cost of the goats (Rs.) | Amount paid |
|---|---|---|---|---|
| 1 | Chinnasangaiya | One | 4500/- | 9 goats were sold and Rs. 1087/- was paid towards loan to the bank. |
| 2 | Paliniswami | One | 4500/- | 5 goats were sold and Rs. 700/- was paid. |
| 3 | Kottai | One | 4500/- | Rs. 700/- was paid |
| 4 | Poonam | One | 4500/- | Rs. 700/- was paid |
| 5 | Alagar | One | 4500/- | Rs. 700/- was paid |

Out of the total 105 sheep 26 were dead, 23 sheep were sold for payment of the loans. Rests of the loan were waived by the Government.

**Electrification:** Electrification is considered as an index of development. Though the Paliyan hamlet in Palamalai was established since the beginning of twentieth century it had not been electrified. The argument was put forth that it was far away from the nearest electrified area, (8-9 K.M.).

Shenbagathope had the electricity facility in its group housing hamlet. This facility was not only used for lighting purpose but it also helped in drinking water supply.

**Medical facilities:** For the medical facility in Shenbagathope a medical van used to come on every Monday with doctors who rendered their services. But after some time the service was discontinued.

A partner of SOCO had opened a medical dispensary with a nurse for the health care of the Paliyans. Unfortunately, the concerned person was busy with his personal work and no nurse was available to stay there permanently. A SOCO worker attends to some urgent medical needs providing first aid. But Paliyans visit this dispensary rarely.

Thus, both the dispensaries are almost defunct at present.

**Other welfare activities:** Along with developmental activities, some other facilities were also extended to the village Shenbagathope or Bhagabati Nagar. Old age pension was provided for three people. In addition they were provided with dhoties and sarees twice a year by the Government and were given noon meal also.

A noon meal centre for the benefit of all the eligible children as well as old age pension beneficiaries had been operating in the school.

Note: Expressing gratitude to Justice Sri P.N. Bhagwati for his sympathy towards Paliyan tribe and interest in their development.

SOCO people renamed the Paliyan hamlet at Shenbagathope as Bhagawati Nagar.

## SUMMARY AND DISCUSSION

The present case study is an unusual kind relating to the upliftment of Paliyans for various reasons. The Paliyans were, till the recent past, food gatherers and hunters with nomadic or semi-nomadic way of life. Their transformation to the life of labourer along with collection of minor forest produce put them into a phase of transition. With the transition during colonial period under the influence of new forest laws, they lost their self-sufficient economic life and became sub-ordinated by the forest contractors. They became labourers under land owing cultivators. In free India when, the Government vigorously tried to abolish the practice of bonded labourers, some leaders tried to emanicipate Paliyan labourers as well.

"The local MLA Tamaraikani started a media compaign to uplift these bonded labourers. Their story came to the notice of the SOCO, a trust dedicated to human rights and social justice." (S. Haja Najeemundeen: Indian Express). SOCO drew the attention of the Supreme Court especially of Justice Bhagabati the then Chief Justice of Supreme court. Finally prominent English dailies of Tamil Nadu disseminated the details of this affair, which made this incident unparallel and thus the efforts to uplift Paliyans and the developmental efforts became unusually conspicuous.

Though there had been efforts for the development of Birhors by Bihar Govt, upliftment of 'Chenchus' by the Andhra govt. the efforts for the upliftment of Paliyans surpassed them all in its media campaign and

**Table No. 6.3 – The summary of the developmental projects and their performance.**

| Sl.No | | Title and sub title | Project proposal | Degree of progress of work | Remarks |
|---|---|---|---|---|---|
| Section | Sub Section | | | | |
| 1 | 2 | 3 | 4 | 5 | 6 |
| 1 | 1.1 | Economic Development<br><br>Co-operative Society<br><br>(Ref- Formation of Co-operative Siceity) | In Order to bring economic development of the Paliyans following projects were taken up.<br><br>Revival of defunct"Shenbagathope Hill Tribals Labour contract co-operative society" Govt. provided Rs. 10,000/- as subsidy towards the capital outlay, Rs. 2500/- as subsidy for purchase of bullock cart, Rs. 4000/- towards cash credit loan | The defunct co-operative society which started in 1965 was revived in 1984, Only tribal people were supposed to become the member while Govt. Officials held the important posts in the society as Ex-officio members. It could not be ascertained how much money Govt. actually provided. | It seems the co-operative society did not work properly. For the first three years members were paid Rs. 6/- per day as their advance wage but did not distribute dividend within three years. Subsequently Govt. prohibited Srivilliputhur range for collection of M.F.P. to save the endangered flying squirrel. Paliyans were unhappy with the functioning of society as they did not receive their dividend in time. |
| | 1.2 | Distribution of mich animals Refer "Distribution of milch animals Refer "Distribution of milch animals | To provide each Paliyans milch buffaloes with a subsidy of Rs. 1250/- (50%) of the cost of Rs. 2500/-. Thus total out lay of Rs. 72500/- was incurred in this scheme. | The buffaloes were selected after the first delivery when the yield of milk is highest. The milch animals were given to selected persons in three stages. Totally 29 buffaloes were given. | Greatest success was in the first year. There was to some extent enthusiasm. In next two or three years and by the end of sixth year all the buffaloes were either sold away or died. |
| | 1.3 | Distribution of goats. (Ref. Goats distribution scheme. | Five people of Shenbagathope were given 5 units (105 goats) at a cost of Rs. 4500/- for each unit. One third of the cost was subsidy. Total outlay was Rs. 22500/- | The expenses were met from IRDP funds. In the first year 26 goats died in normal course 23 goats were sold, each beneficiary selling the number of goats as per his convenience each paid Rs. 700/- at the 1st instance excepting one who paid Rs. 1087/-. It seems rest of the money was waived by Govt. | The objective was to provide a regular source of income through goat rearing. But the entire 105 goats were exhausted either through sale (around 25%), consumption or due to natural death. Thus the beneficiaries got some benefits but the objective failed almost totally. |

| | | | | |
|---|---|---|---|---|
| 1.4(a) | Collection of medicinal herbs. | Ayurvedic manufacturing institute TAMPCOL started collecting medicinal herbs through Paliyans of Shendbagathope for their economic development.<br><br>Neither the Govt. nor TAMCOL had to spend in training Paliyans. | TAMPCOL was successful in motivating some people to collect medicinal herbs, since some people had already used to do such work, they could earn a hand some amount through such work. | Since this type of work is usually done by the Paliyans they got further encouragement when officials came forward to persuade them to do such work. The payment of wages was also fairly attractive. The project was fairly successful. |
| 1.4 (b) | Cultivation of medicinal herbs. | It was propsed to take 33 acres of land on lease from Arulmigu Andal temple 25 Paliyans were to be engaged on daily wage basis. | In order to keep implements and stock of herbs and for the stay or officials the teachers quarter was given for the time being until the old one was repaired. | There was no sign of the cultivation of herbs during the field work of the researcher. |
| 1.5 | Distribution of land among the Paliyans Ref- Land Assignment. | a) 293 acres of forest 'Poromboke' land proposed in Mettupali village in which 1955 trees to be cut and utilized by the match industries.<br>b) Industries besides forest poromboke lands measuring 61.56 acres had been selected for assignments.<br><br>Estimate of 4.75 was suggested for reclaiming land. | Since the Poromboke land of 61.56 acres is not suitable for cultivation the collector of Ramnathpuram had sent an estimate of Rs. 4.75 lakhs for reclaiming the same land to make the land suitable for cultivation | There was no progress either in getting the land on lease basis or on transfer to Paliyan beneficiaries. Also no progress in reclaiming the land was done for cultivation. Thus the proposal was a total failure. |
| 1.6 | Providing employment opportunity for the Paliyans. | 30 labourers were proposed to be recruited to work in Poovani orchard | No calculation of financial out lay There was no keen interest shown by Govt. for employment of the Paliyans. | It was a grand failure. There was no effort to implement the proposal. Neither the Paliyans were interested to work as day labourer in the Poovani orchard. |
| 1.7 | Providing employment through Khadi and village industries | | | |

| | | | | | |
|---|---|---|---|---|---|
| | 1.7.1 | <u>Honey extraction</u>. The training in bee keeping industry | Since honey collection is an important traditional occupation of Paliyans emphasis on honey extraction was given financial outlay of Rs. 22500/- for imparting training in bee-keepi8ng, cost of materials and two bee-hives cost including, total outlay was Rs. 41200/- | Under Integrated Rural Development programme 30 people were trained in beekeeping for a period of two months. Each trainee was given two bee-hives. Training cost of Rs. 6500/- was provided from the TRYCEM funds. Training started from 1.2.84 | In Hindu, on 12 the October 1984 under the caption "Enabling them to lead a better life" a picture shows how" a tribal youth who has set up a bee-keeping unit." But during field Investigation a single person was not found continuing the bee-keeping |
| | 1.7.2 | Training in basket making for women | For women's employment it was decided to train them in manufacturing house hold goods 20 women were trained in basket making. Training was for 6 months with a stipend of Rs. 125/- per month. Total expenditure was Rs. 19000/- training stipend was Rs. 15000/-. Purchase of tools at the rate of Rs. 200/- each amounting to Rs. 4,000/-. Total outlay was Rs. 19000/- | Twenty women were trained in basket making for six months at monthly stipend of Rs. 125/- each.<br>After the manufacturing of baskets these could not be sold. | Training was a good thing. But after training the trainees could earn only Rs. 2/- per day. Perhaps they could not compete with the professional crafts men. Neither multipurpose society fulfilled its obligation in marketing nor the trained women were serious in pursuing the craft in which they were trained. |
| | 1.8 | Training in power Loom | To train weaving in Power loom. Adi Drould and Tribal welfare Deptt. Provided financial support for the training | 5 youths of Shenbagathope were trained inweaving in a power loom at Srivilliputhur. | The candidates attended training without any regularity. None has ever earned through weaving with a power loom. |
| | 2.1 | Improving the working condition of the school. | 1. Electrification<br>2. Provision of a quarter for the teacher.<br>3. Noon meal centre for school children.<br>4. Repairing of the school building | All the four proposals were fulfilled by the Govt. | Though all the four proposals were fulfilled it did not improve the teaching in school. No teacher stayed in his quarter and Govt. gave orders for utilization of the vacant quarter for other purpose than improving teaching in the school. |
| 3 | | Improving health condition of the Paliyans | | | |
| | 3.1 | (Ref-Mobile Health Camp.) | To provide medical facilities | A mobile medical team had been formed and they looked after the medical needs of the Paliyans of Shenbagathope. The mobile medical team used to visit the village every Monday and Friday. | The medical unit worked for nearly a year but later it was discontinued. So this short term effort did not improve health condition substantially. |

| | | | | |
|---|---|---|---|---|
| 4 | Road communication facility | | | |
| 4.1(a) | Improving the road to Srivilliputur | Out of the road of 10 km. from Srivilliputur to Shenbagathope, 6km road tobe black topped | The 10km Read was constructed earlier and 6 km. Black topping was taken up urgently. | The project was completed and was successful in improving the Paliyan's communication |
| 4.1(b) | Providing bus facility Srivilliputhur to Shenbagathope. | Town bus for two trips one in the morinig and the other in the evening from Srivilliputhur to Shenbagathope and back were proposed. | Both the proposals were speedly completed. | It brought tremendous change in the life style of the Paliyans and removed their physical and social isolation. |
| 5 | Provision of accommodation | | | |
| 5.1 | Group housing scheme (Ref. Chapter – V) | It seems the group housing scheme was implemented in Shenbatathope not as a special developmental effort but as a general plan introduced by Govt. of India. | As discussed earlier in Ch-V the first group houses started in 1965. The second and third phase houses were started after abolition of bonded labour movement. | In Chapter – V how second phase houses were successful and how third phase houses were shortly damaged had been discussed. |
| 6 | Provisions for essential goods | | | |
| 6.1 | Supply of essential commodities. | In the absence of grocery store in the village, Govt. planned to start a fair price shop in the village for benefit of Paliyans. | Fair price shop was started in Feb-1984. In a mobile van. All essential commodities such as rice, dal, Kerosine, palmolin Oil etc. used to be brought to the village along with other controlled commodities. | There scheme was very successful and people were satisfied with this operation. |
| 6.2 | Issue of family cards | A proposal was made to provide family card to each family for supply of rice etc. | The proposal was sent to commissioner of Civil Supply. | There was no response to the proposal and hence a failure. |
| 7 | Misc items | | | |
| 7.1 | Old age pension | Old age pension (OAP). Along with other developmental facilities Govt. officials sanctioned old age pensions for the welfare of the elderly Paliyans. Three people were benefitted. | Three people were granted old age pension and were entitled to get mid day meal. | The people who got old age pension hardly take mid day meal. They live in Medu, far away from the village. Hence they could not come to take meals. The informants said that any other person of the village or any other guest could get that mid-day meal. |

swiftness. Here, the unprecedented sympathy of Chief Justice of Supreme court, vigorous coordination by the collector of Madurai representing the Government of Tamil Nadu, the careful monitoring by SOCO, Trust, made the effort of uplifting Paliyans most outstanding. Finally the media coverage with colourful titles made the details and stages of development immensely attractive. The details of different developmental projects are presented in summarized form in the table below.

The Paliyans of Shenbagathope live in plain area at a little distance from the foot of a hill with a segment on the hill slope. In such sparsely populated foot hills the collection of MFP is the principal occupation even today, hunting and food gathering for subsistence being as an auxiliary activity. They had their self sufficient economic life but now the equilibrium has been lost owing to invasion of migrants, forest control and commercial use of resources around them.

Rather being unable to take advantage of the opportunities available around them due to such change owing to their illiteracy, ignorance, simplicityand shyness, they have become victims of exploitation by such affluent people who migrated to these hill areas in search of prosperous fortune. For food they could manage with their food gathering habits but as the socioeconomic needs increased they had to depend on outside market which determines the cost of their M.F.P. and the crafty business men have upper hand to fix the cost of their goods which they collect with such strain and endurance. Earlier members of the single Nadar family could determine the cost of such MFP since their main occupation is collection of minor forest produce. Paliyans depend solely on the persons who take lease of the "Andal Koil tope" (estate of Andal temple) and land (with an area of 250 acres). The other important man is the person who gets the right to purchase all the minor forest produce.

When the co-operative was revived, District Collector assured that all the forest beat would be given to the Paliyans for collection of MFP. That inspired some Paliyans to join the cooperative society but what happened during the early stage of the co-operative was not very clear. It seems (as published in The Hindu Dated the 26th June'1987) the right to collect and sell of MFP was controlled by forest department. Co-operative society was not given any right although the Paliyans were ready to Pay Rs. 7000/- for a single beat of the Alaghar koil area alone. On the contrary all the beats within the Srivilliputhur forest range were given at a cost of Rs. 16,000/ to an outsider. In spite of earlier decision to award the beat to the newly formed cooperative society exclusively run for the benefits of the Paliyans, the action was not on the decided line.

At this stage another incident occurred. As mentioned earlier, the flying squirrel which was a rare and endangended species, was to be protected. So Govt. reserved the entire forest area of Srivilliputur range and collection of MFP was totally prohibited. The main food item of the flying squirrel is tamaring and therefore collection of tamarind was restricted. But the forest officials took a liberal stand on the issue based on Forest Act 1988 and allowed the Paliyans to collect MFP. The Nadar business family who was controlling the sale of MFP adopted a tactics. At this stage the child who became famous during the enquiry on bonded labour became the prominent head of the village. Now Nadar kept him in his control and even utilized police in harassing him so that he could keep him in his clutches.

Earlier the godown of the Nadar was in Srivilliputhur and the Paliyan collectors had to carry MFPs to the godown at Srivilliputhur. He provided accommodation in his house & entertained the youths by showing films / blue films. Even foreign liquor was provided for these Paliyans and thus what they earned by selling MFP got dwindled by such costly entertainments. According to the Nadar the Paliyans are extremely lazy and once they acquired surplus money they would rather sleep than earn money for their economic well being. Thus by making Paliyan youths addicted to liquor the Nadar trader could keep them in his clutches to his advantage. In the circumstances when both the lender and loanee (creditor and debtor) are reluctant to express real facts, I could not collect information whether the Paliyans are indebted to the Nadar or not, if so what was the loan amount borrowed by each of them. For this I dropped the idea of making any further queries. Yet it is clear that inspite of all the din and bustle, the Paliyans still remain subordinate to Nadars or other MFP contractors and have to depend heavily on them whose wealth is generated through the exploitation of the Paliyans. Had the Shenbagathope forest labour co-operative been functioning properly it had the scope to accommodate the poor Paliyans on a permanent economic security base. But due to negligence or mismanagement of the official executives and the unofficial managers the co-operative could not achieve the expected success.

Next to the Labour Co-operative which in the beginning phase procured and sold all the MFP, the project that could have emancipate the Paliyans was the "Dairy development cooperative" through which the distribution of milch buffaloes was made. Here more labourious beneficiaries could repay, within one year or ten months, the total subsidzed loan amount beside leaving a margin earning of Rs. 25/- or more per week. Here again it was the failure of the Government to provide cattle feed which could increase milk amount substantially. Finally people represented to the Government to provide shelter for their cattle but latter could not help substantially except giving an order that the old buildings constructed by Adi Dravidar Department which were under repair, could be used for the accommodation of the buffaloes. The worst thing was that instead of guiding properly how to save money and feed the cattle when it does not yield any milk, the Government officials and even some social workers advised the Paliyans, it is reported, to sell away the cattle and repay the balance loan amount. The Paliyans were very enthusiastic in the early years and it was told that even old people and women were very busy in grazing cattle. Besides, the local surrounding caste people must have been very much allured by the attractive subsidized cost of the buffaloes and must have lured the Paliyans to sell their buffaloes. Thus a very perspective programme with a great potentiality for the success was completely nullified due to unscrupulous Government officials, greedy jealous and rich neighbours and a few poorly paid social workers.

The next alluring programme was to provide land to the semi-nomadic Paliyans to lead a settled life through agriculture. Though several spots of forest land could be suggested for distribution, a huge amount of Rs. 4 lakhs was required for reclamation of the forest and the Government could not provide the amount speedily.

Besides the Government programme to train both men and women in bee keeping basket making and weaving could not yield success either due to lack of finance during the gestation period or for failing to compete in marketing the goods produced by the newly trained people. Other programmes such as mobile medical unit, and bi-monthly fair price store had their natural death due to lack of regular or permanent provision of funds. Further the NGO which brought the miserable condition of the Paliyans to the notice of the Government could not build any sustainable development programme. Finally the Paliyans of the village lost interest and enthusiasm when the potential programmes failed due to lack of management skill and absence of moral principles of government officials, the NGO's and its social workers. Thus inspite of well publicized programmes of the government and the SOCO, (the NGO organization), no sustained development could be achieved in the poverty stricken Paliyan society. It requires more detailed and careful study of developmental efforts which were initiated by the Government and the SOCO, the voluntary organization. The efforts should be made to analyse why these projects failed inspite of government co-operaion and peoples willingness to adopt several innovations that could have brought substantial improvement in the financial well-being of the Paliyans. The factors especially social factors that brought the total failure of the developmental efforts aimed at bringing socio-economic security to the poor Paliyans are required to be analyzed and projected. We have tried to show this in conclusion. Finally further discussion has been taken up regarding the role of different agencies in the developmental process and their success failure in the concluding chapter.

# Conclusion

The present book "The Ecological and Economic Contexts of Paliyan Development" is meant mainly to present the developmental process amongst the Paliyans. The chapters contained in the book have been broadly framed into two parts. The first part consists of the cultural background of the Paliyans. It describes how their culture core helped in developing their subsistence pattern, economic life and social structure by adopting their life to the environement of their habitat. Subsequently the culture contact situation affected both ecology and economy and gave birth to various problems due to exploitation from different quarters. In the second part developmental process amongst the Paliyans in the background of their changing economy and development have been presented. Both these parts have been covered in six chapters of this book. The observations on the present study are as follows.

The Paliyan culture appears to be in a process of rapid change perceived to be commenced from the colonial period and accelerated during post independence period. Their nomadic habit has been reduced. Their number has increased demographically and their homogenous culture based on natural environment has been diversified due to external intervening variables. Some of the acculturised Paliyans adopted sedentary life styles.

In pre-colonial period Paliyans had their roots in the natural forest environment which influenced their culture core to a great extent. For some Paliyans the vanishing forest replaced by commercial plantation and estates and the changed natural environment modified their cultural behavior substantially with inter-cultural pressure being a prominent intervening variable. In some habitats their culture is undistinguishable from that of neighbouring people. Paliyans continuing in their familiar natural environment have retained many traits of their original culture.

Paliyans appear to have undergone two cycles of acculturation process. The first one, perceived to be some five hundred years ago did not disturb their traditional way of life, rather they adopted Sanskritization by adopting other caste usages.

Second acculturation process started within colonial period and continuing till date. This substantially modified their traditional culture due to change in ecology and economy.

Prominent changes in their settlement, housing, dress and other items of material culture such as use of cooking pots, utensils, bows and arrows or borrowed guns have been observed. Rice and other food items available in the market are their present day daily food. Previously nearby forests were having abundant diascorea, green vegetables and other varieties of food. Either individually or with domestic labour group they were collecting such food materials. But now with emphasis on collection of MFP and working as labourers bring direct cash by which the food gathering habit has been reduced to a great extent. Much practiced hunting have been reduced to a ritual or leisure activity.

Foraging has been modified by agricultural labour, collection of MFP services in organized and unorganized sectors and other types of labour.

Previously all Paliyans were following great and little tradition of Hindu religion, but Christianity is now making a dent in both northern parts (Palani hills) and extreme southern parts of Tirunelvelli district.

Paliyans in general have diversified and different groups are in different stages of acculturation and progress. Being an ethnic minority group, Paliyans have passed through at least three phases of acculturation i.e. as isolated tribal groups, semi isolated tribal groups and now in permanent contact. But still they are far from fully integrated phase.

Paliyans earlier attitude of shyness and avoidance of outside people has been influenced by wider culture contact situation and intercultural interventions. Earlier non-aggressive nature, intra-group cordiality timidity and shyness are reducing.

Social customs like marriage, child rearing practices have been changed. There are cases of inter caste marriages being tolerated excepting with lower castes.

1. Through intervention of Governmental and Non-Government Organizations' activities modern allopathic medicines have made its way into traditional therapeutic practices.
2. The economy of the Paliyans is slowly moving from self sufficiency subsistence to monetized market economy.
3. Due to new forest laws during colonial administration Paliyans are restricted to pursue their traditional occupation. They had no

resources to adopt any alternative ones except selling their own labour, to forest contractors land owing cultivators and plantation owners.

4. At least three behaviourial factors can be altributed to present exploitation of Paliyans, ignorance, absence of overt aggression and shyness.
5. They are unable to understand and adapt to today's monetized market economy and most often turn victims to the unscrupulous business men and tricky employers.
6. Present Paliyans are attracted by the modern dress and other attractive commodities where they are also sometimes cheated during transactions.

## Observations on developmental process among Paliyans

Tribes of peninsular India are somewhat different from central Indian tribes. A few of them including the Paliyans were numerically small groups and lead a seminomadic, band life. Neither they had clan organization nor wide ranging kinship organization. They had developed a non-aggressive personality and more or less maintained cordial relationship with their neighbourers.

When Paliyans were alinated from their own habitat through planned forest management, commercial plantation and utilization of the forest areas etc, they became an unprotected community exposed to crucial monetized world outside. The land revenue system introduced during colonial era acted against their interest. They became landless due to their nomadic characteristics and were dispossessed of their own forest reserves upon which they had built their subsistence pattern and culture from time immemorial. With changing ecology when they lost the scope of following their traditional economic life they had to adopt the life of an exploited labourer.

Government sponsored developmental activities intended for their uplifement akin to other tribals of the country did have mixed results, the failure results being more prominent in this community. It is difficult to suspect the motive of Government but the activities demonstrate lack of sincerity and synchronism among different organs of government.

Development does not imply economical upliftment alone but includes better quality of life as well. Indices include education (literacy rate), health, (infant mortality rate) women's development, life expectancy and so on.

Sustainable development refers to constant improvement in standards of living overcoming all resistances, including unforeseen factors.

Case studies conducted at the sample villages bring various factors into focus while reinforcing some such as transfer of macro level planning to grass root level, endogenic changes for sustainable development, government's role and Paliyan's absorbing capacity, inter cultural and other social interventions impeding the developmental processes. Macro level planning formulated at top echelon of the bureaucracy or at centre far away from the implementation site lacks first hand experience and understanding of the problems. At times it either ignores certain factors or over emphasizes some depending upon the biased reporting from bottom rungs of administration. Threrfore, macro level planning of the tribal development visualized for entire tribal population needs to be realigned when these are to be implemented at micro level. Since each tribal group is different in their characteristic, nature, habit, geographical environment, demography and culture etc. such even application of programmes does not create desired results. Each community's customs need to be analysed for their socio-anthropological adaptability of the schemes.

The case of bee keeping and basket making schemes may be a case for the issue. Paliyans were efficient honey collectors but they neither had the acumen nor the attitude for rearing bees which required close nursing and understanding their behavior. They could weave a basket but had little knowledge in marketing. Regular income required sustained marketing abilities even beyond local markets which Paliyans lacked. Government could have helped them on this to succeed.

Implementation of development activities in the tribal habitats like Paliyans require at least three eligible agents – the benefactor, beneficiary and the domain.

Paliyans as the beneficiary of the developmental schemes implemented either by Government or by N.G.O's lacked many vital prerequisites crucial to their upliftment. They are innocent, simple, illiterate and shy which lead to their exploitation. Being partly nomadic they lacked, stable community organizations, strong leadership pattern and lacked endogenic changes for development prospective attitude. As a result they were unable to understand and comprehend that the developmental programs were intended for their benefit and their active support for its implementation is crucial for upliftment. Even though they have travelled from seminomadic to permanent contact situation of acculturation process they are not attached to any particular location and development of physical facilities like houses roads etc are least attractive for them. At least one factor could have induced them for cooperating in the process.

The developmental projects could have easily classified into two types (i) Asset building and (ii) Income generation. Asset building activities like allotment of agricultural lands or land reclaimed from the forest could have allured them to an organized settled community life through which other income generating activities could have been implemented and monitored.

But Governments intention of providing lands to each of the families got rotton through bureaucratic process. Paliyans could have been included as the displaced categories due to forest reservation which could have entitled them for certain resettlement privileges. But being nomadic, bureaucratic confusion and apathy resulted in their distress which needs earliest rectification.

Paliyans lack prospective attitude of life and subsisted on daily collection of food or wages. They do not understand the advantage of saving or implication of infrastructural development to their advantage. This has resulted in apathetic manners to developmental programs and projects.

Income generating developmental projects / programs with high gestation period is bound to frustrate Paliyans since their life depend upon daily collection and income. Any occasional extra income is misutillsed in cheap entertainment and does not necessarily lead to tangible asset creation or savings.

While government as go benefactor initiated every effort for the upliftment of Paliyans, its intentions and concepts at a national level are not fully understood by the workers and officials implementing it at lowest rung. They lack sufficient motivation and spirit. As a result the presence and contact with tribals for the schemes are of shorter duration and does not create necessary effect on sustainable basis. This is more prominent in the case study conducted on Paliyans. They spent less time with beneficiaries, induced less, explaining the programs. Their interest could have been observed through creation of adequate local organizational structure and development of indigenous leadership pattern to bring sustainable endogenic change of culture and development. This would have helped them to pursue the programs even after withdrawal of external agents. NGO involved in the developmental activities pursued this approach to some extent, but withdrew shortly after.

Lack of inter departmental, coordination between various arms of Government is clearly visible in case of Paliyans, study. While Collector agreed to provide facilities required for Paliyans, other departments like forest, agriculture, cooperative etc involved in the activity failed to honour and implement the commitments.

Post scheme evaluation of an activity formed another prominent lapse in Paliyan's development. Government not necessarily sought honest feedback on the implemented program. A scheme under implementation required some time to sink in and create visible effects. It requires frequent visits and motivation and monitoring to assess the effect and trace possible causes. It require constant evaluation and correct impartial observation which is to be fed back to bring changes. Modification and implementation could have visible effects which has been ignored in the Paliyan's development.

Domain of the activity refers to physical and socio cultural settings of the tribal people. Each community is distinguishable in their characteristics. Domain is the medium and surroundings in which programs are implemented and provides a favourable environment to friendly programs. Small ethnic groups like Paliyan's never influence or control their settings but rather influenced by it. Till date they have not developed the habit of organizing themselves politically to work as a pressure group to compel government to provide funds and infrastructure for their socio economic development. Nor they dominantly influence other caste people to have economical advantage. Rather they are controlled by other's perception and attitude and this has an effect on Paliyan's development.

Comparison of developmental activities at the two distinguishable location reinforces the argument that strong community leadership and community organizational structure in small ethnic groups can help sustainable development.

Empowering target tribal groups like Paliyan's in the developmental decisions to solve their own problems is to be recognized in spite of lapses associated with it. Government officials fearful of the mismanagement or attracted by the authority never divest the decision talking. The cooperative society is to collect and sale MFPS, was formed with non-tribals forming a majority. As a result the Paliyan members did not get scope to control the affairs nor influence the Government. Cooperative Departments did not care and the forest department diplomatically avoided awarding MFP collection contract to the society. As a result a nice feasible venture collapsed.

The case study still brings another aspect of development into limelight. The reaction and attitude of neighbouring caste people is extremely interesting and with careful investigation can highlight the shortcomings in the implementation of the project.

Milch animal scheme is the best planned project for long term poverty eradication of the Paliyans. In the first year majority could repay their loans and some even took loans for second time. The scheme appears to be successful and projects no procedural deficiencies, yet after sometime the Paliyans sold all their buffaloes to their neighbours. Investigation revealed biased advice by neighbours and officials. Paliyans neighbouring communities enjoyed and availed their cheap and easily available labour. When Paliyans showed income from sale of milk and repaid all their loans, this did not appease neighbours. They always depend upon latter's advice very much. Paliyan's prosperity attracted neighbours covetousness and they advised former to sell their buffaloes. Official corroborated and the scheme failed.

The developmental activities pursued in both the areas have differential results due to ecological factors. The village Palamalai is situated in a fertile valley where commercial crops are grown and some of the Paliyans grow these by taking lease of 'Porombok' land. But in Shenbagathope though there are plenty of plain land they are owned and cultivated by non tribal people. Due to vast forest resources the people of Shenbagathope village sometimes earn fairly good amount of money by collecting MFP. But since such people are exploited heavily and are used to lead a prodigal life or do extravagance of expenditure they remain poor as other tribals while some of the villagers of Palamalai can lead a descent life due to their frugal and controlled style of life. In case of villagers of Palamalai they get proper advice from the Father while in Shenbagathope villagers are guided by forest contractors and traders who exploit them in subtle ways.

The developmental process can not be pursued in an isolated manner only for a small community. No one lives in isolation. In India though each caste or community has its distinct identity there is interdependence of each group with others. If the Government officials or NGO workers concentrate on one group, other groups try to non-co-operate and try to misguide the favoured community and then nullify the development efforts. Therefore the persons who are trying to take the help of a particular community should try to help other castes and communities so that all can make a cohesive attempt to achieve success in the developmental efforts. In a paper 'Tradition and Development in tribal Orissa. A case study of Dongaria Kondha Development Agency" Nayak refers how amongst the Dongria Kondhs of Orissa several Government officials had worked. They isolated Dombs (S.C.) and tried to implement development projects for Dongaria Kondhs without having any contact with Dombs who exploit the Kondhs. All such officers failed to create sustained development. One young anthropologist who was employed in Dongaria Kandha area tried to help the Dombs for their development and there was greater success in the development of both Dungaria Kondhs along with development of the Dombs. *(Nayak: 1988: 82 ; 85).*

India is the biggest democratic nation in the whole world. On the contrary its social structure built on caste system is basically hierarchical in nature. In the day-to-day life, members of each caste are conscious of their status and expect privileges associated with their caste occupation inspite of the fact that sometimes it violates the democratic norms. This phenomenon sometimes leads to conflicting situation and negates the successful implementation of developmental projects.

In order to bring sustained development, the first priority is to create leadership pattern in the target group who can disseminate the information regarding development and can identify the elements which stand as barrier to the developmental process. The success of any developmental project depends on such leaders who can give guidance to their community members. But creating such leadership pattern is not an easy affair. In the present case SOCO tried to build some leaders among Paliyans at Shenbagathope but their efforts were short lived. They failed for lack of characteristics and without giving them adequate training it was not possible to build effective leaders who could give beacon light to their community members. Selecting young people as leader may not be always effective to give right decision. In many societies council of elders play very important role in giving guidance to their community. In that light, council of elders can be constituted and younger people should be encouraged to work with such council of elders. This is no doubt a difficult task. But building such a council of elders and developing a youth group to work with them can build up more effective leadership pattern, who can bring sustained development with proper and adequate training and guidance.

We have discussed above that for successful implementation of developmental projects many agencies could work together. Out of them first position should be given to the community itself or its leaders who can be utilized for the development of the total community. The next position automatically goes to government who has the rights and authority to implement developmental projects as its basic or primary responsibility. Government is no doubt very powerful and has both the human and financial resources to carry out developmental activities. Government organization contains the best brains of the society and can chalk out very valuable and pragmatic plans. In Indian society Government Officials always enjoy a high social status. They most often can not go down to the level of down-trodden people. Many of the officials can not discharge their duties very effectively as social service personnel. But government people are very rigid so far money matter is concerned. Thus they lack dynamism. On the other hand personnel of voluntary organization are comparatively having more dynamism and each has to take independent decision on crucial issues. Yet unfortunately most of the voluntary organizations face the financial problems and fail to act properly. Some corrupt voluntary organization fail to play their proper role.

Bunker Ray observes, there are lessons to be learnt in how voluntary agencies have demystified technology; how they have managed to delink educational qualification and experience, responsibilities, intelligence, aptitude and competence; how organization have grown without becoming institutions, how they have used the system without succumbing to it and loosing their identity; how they have spread their radical ideas and made them acceptable to government to try them out on a larger scale; how within an organization it has been possible to decentralize, to delegate and handover to the people in the area. (*Sundaram: 1986: 161-162*).

However efficient a Government official or a NGO official may be the community members or its popular leaders are the right persons to participate in the policy making process and in implementing the developmental projects. In any case the agencies to which category they may belong to it is obvious that they should keep constant contact with the target group and should convince the people how the implementation of developmental projects are beneficial to individuals and the community.

The development of the Paliyans can be based on alteast three factors. (1) Education (2) Information and (3) Participation. The development is a process of change and it requires change of traditional way of life, adoption of innovation which are related to new aspiration and ideologies that can transform orthodox way of life and give inspiration for adopting new technology, organization that can improve socio-economic life and that can bring greater prosperity and also can bring a higher quality of life.

Education is almost indispensable for instilling new aspiraition for adoption of essential innovations that are vital for generating inspiration for development.

In a developmental process education does not merely mean the learning of three R's (Reading, Writing, Arithmetics). Education which can generate a new out look towards life which can equip a person for substantial earning to meet basic necessities of life and free him from poverty jealousy and meanness. Acquiring such knowledge is difficult but improvement process is the vital need for tribal development. Formal and Nonformal education can be reorganized and transmitted to Paliyan youths who can really help to generate an environment for development in future. Not only contexts but also transmission methods need to suit common tribals like Paliyans.

Information forms another core activity of development. To generate awareness and instill transparency in activity which in trun will induce acceptance, adequate information on the plans and programs is a vital necessity. This will reduce exploitation, assertion of the rights and privileges available as a citizen and instill confidence in simple Paliyans suffering from inferiority complex. Decision making and choice will improve, can help quality of life. The medium of information need to be carefully chosen to be acceptable to the Paliyans.

Participation of the target community in the developmental activities has been widely propagated concept of the social anthropologists.

The developmental strategy at the present context is implemented by members of bureaucracy or by members of voluntary organizations. Unfortunately in most of the projects for development the planner take all precaution but miss the target group for whose benefit the entire project is planned. In any developmental process it is the people who matter first. The people's participation in the planning stage and in implementing the project is crucial. Their co-operation and participation, being the beneficiary of the developmental activities, and cohesive efforts to create the situation of development can not be over emphasized.

# Glossary

*Akka* - Elder Sister
*Amma* - Mother
*Aurval* - Bill hook
*Chedi* - Plant
*Irumal* - Cold
*Kadu/Kavu* - Forest
*Kallu* - Stone
Kalyana Mandap
- Marriage hall
*Kirai* - Greens
*Kilangu* - Tuber
*Kodi* - Creeper
Kovil/Koil - Temple
*Kottai* - Seeds
*Kombu* - Stick
*Madu* - Cattle
*Maram* - Tree
*Medu* - A land at a higher level
*Onu* - One
*Pal* - Milk
*Palam* - Fruit
*Pallu* - Teeth
*Parai* - It is portion of a mountain
*Pattai* - Bark
*Porombok* Land- Government Land
Pucca Building - Permanent Building
*Pusini Kai* - Pumpkin
*Raja* - King
*Sengal* - Brick
*Sondakar* - Relatives
*Sunambu* - Lime
*Uppu* - Salt
*Vaita* -Stomach
*Vandi* - Vehicle
*Ver* - Root

# Appendix

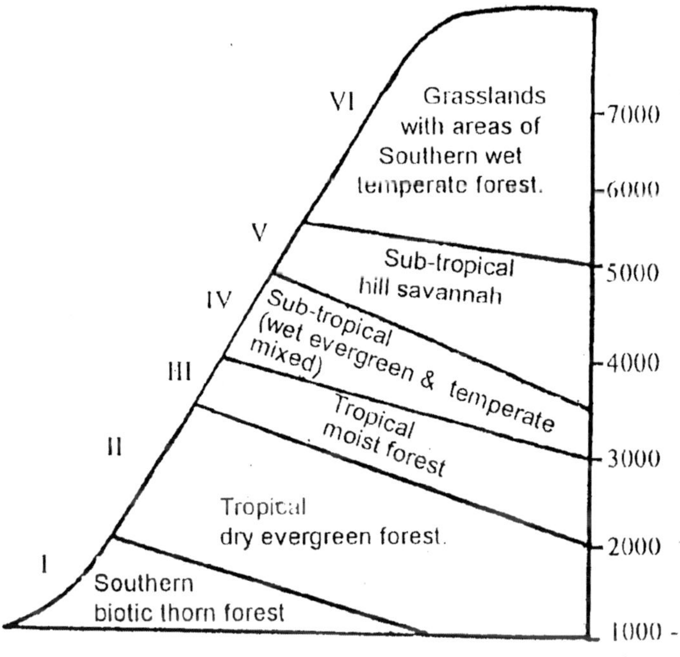


Zones of forestation (I-VI) in Paliyan occupied ranges.
Derived mainly from Puri, 1960 and Gardner, P.M. 1965

# Bibliography

**Aiyappan, A**

1948 *Report on Socio-economic conditions of Aboriginal Tribes* of the Province of Madras, Government Press.

1980 The Situation in the South. Published in the *Illustrated Weekly* special issue "Tribals of India" Vol.c.144, December 14-20. Pp 22-25.

**Aluvalar Manamakil Mondram**

1992 *'Our Kodaikanal'*

**Arnold David, Ramachandra Guha (ed.)**

1995 *Nature Culture Imperialism:* Essays on the *Environmental History of South Asia*. Oxford University Press, Delhi.

**Balija, B.S.**

1960 *Madras District Gazetteers:* Madurai, Madras: Government Press.

**Barth F.**

1956 "Ecological Relationships of Ethnic Groups in Swat, North Pakistan" *American Anthropology* 58: 1079-89.

**Behera, D.K.**

1993 "Impact of deforestation on the life of the Plain Bhuiyans". *Journal of Human Ecology. 4(3).*

1996 Development, Environment and the People,' An anthropological Perspective *South Asian Anthropologists* Vol. 17 1996 No. 2 Sarat Chandra Roy Institute of Anthropological Studies, Ranchi.

**Behura N.K.**

1995 'Tribes of India: Planned Development' In A.K. Singh and M.K Jabbi (ed) Tribals in India: *Development, Deprivation and Discontent*. Har Anand Publication, New Delhi.

**Bhat C.P.**

1995 "FORWARD" *In All About Draft Forest Bill and Forest Lands.* Hiremath S.R. etal (ed) Samaj Parivartan Samudaya Dharwad and Centre for Tribal Conscientizatin, Pune etal.

**Bicchieri M.G.**

1972 *Hunters and Gatherers Today: A Socioeconomic Study of Eleven Such Cultures in the Twentieth Century.* Holt, Rinehart and Winston, Inc. New York.

1990 'From Eurocentric to Geocentric the perspectives from small scale societies' *Anthropos 85.*

**Bird-David, Nurit H.**

1988 'Hunters and Gatherers and other people – a reexamination' In hunters and gatherers 1. History, evolution *and social change.* Tim Ingold, David Riches, and James Woodburn, ed. Oxford: Berg Publishers Limited.

**Borian M.P**

*The travail of tribal workers* – (unpublished paper) Gandhigram, Tamil Nadu

**Brandis, D**

1897 *Indian Forestry*, (working plan)

**Burch, Ernest S. Jr., Linda J. Ellanna (ed.)**

1994 *Key Issues in Hunter – Gatherer Research*, BERG Oxford / Province, London.

**Caritas**

1951 The late Joseph Sanniasiar and the Paliyans of Palamalai.

**Cernea Michael M.**

1990 'Putting People First; Social Science Knowledge for Development Intervention' in Hari Mohan Mathur (ed) *The Human Dimension of Development Perspectives from Anthropology*, Concept Publishing Company, New Delhi.

**Chaudhuri Buddhadeb (ed.)**

1982 *Tribal Development in India: Problems and Prospects.* Inter-India Publication, Delhi.

**Dahmen Rev. F.**

1908 'The Paliyans: A Hill Tribe of Palnihills' *Anthropos 3.*

**Dasgupta Samira**

1994 *'Birjia: Society and culture'* Firma KLM Private Limited, Calcutta.

**Desai A.R.**

1987 *Social background of Indian Nationalism* (First published in 1948). Popular Prakashan.

**Dikshit K.R.**

1991 *Environment, Forest Ecology and Man in the Western hats: The case of Mahabaleshwar Plateau. Rawat Publications, Jaipur*

**D' Penha, G.F. 'The life of the Paliayans'**

1902 Indian Anthropology

**Dube S.C.**

1982 'Unrealized Utopias and Widening Disparities'. In L.P.Vidyarthi (ed.) *Rural Development in South India*. Concept Publishing Company, New Delhi.

1992 *Understanding Change: Anthropological & Sociological* Perspectives. Vikas Publishing House Pvt. Ltd. New Delhi.

**Dumont Louis**

1966 *Homo Hierarchicus: The Caste System and its Implications*. Vikas Publication New Delhi.

**Elwin Verrier**

1963 *'A new Deal for Tribal India'* Ministry of Home Affairs, New Delhi.

**Fazl Abul**

1873 *Ain-I-Akbari*, H. Biochmann, Calcutta Vol I-II

**Fernandes, W.G. Menon and P Viagas**

1987 *Forests, environment and the Forest Dweller Economy in Orissa*, Indian Social Institute: New Delhi.

**Fernandes Walter**

1990 Forest Policy: A Solution to Tribal Deprivation. *The Indian Journal of Social Work Vol – L-1, No.-1.*

**First Enquiry Report**

1983 Please refer to Gurumurthi 1983

**Foster, G.M.**

1973 *Traditional Societies and technological Change, Calcutta:* Allied Publishers.

**Foulkes. R.**

1932 'A Note on the Paliyans of Madura District' in Yeatts, *Census of India* 1931, Vol. 14, Madras.

**Fox, R.G.**

1969 *'Professional Primitives:* Hunters and Gatherers of Nuclear S. *Asia'* *Man in India,* Vol. 49, No.2

**Francis W.**

1906 Extract from *The District Gazetier of Madurai* Vol-I

**Gadgil M & R Guha**

1992 *Fissured Land, An Ecological History of India* Oxford University Press.

**Gadgil, M & K.C. Malhotra**

1994 'The Ecological Significance of Caste' in R.C. Guha (ed), *Social ecology*, Oxford University Press, New Delhi.

**Gardner P.M.**

1965 *Ecology and Social Structure in Refugee Population:* The Paliyans of South India. A dissertation in Anthropology, University of Pensylvnia.

1972 '*The Paliyans. In Hunters and Gtherers Today*, M.G. Bicchieri, ed. Holt, Rinehart and Winston, New York.

1982 'Ascribed Austority ; A tribal Path to Purity' *Man 17.*

1985 'Bicultural Oscillation as a long term Adaptation to Cultural Frontiers'. *Human Ecology 13.*

1988 Pressures for Tamil propriety in Paliyan Social Organization. *In Hunter's and Gatherers.* Vol-1: History, Evolution, and Social

Charge. T. Ingold, D. Riches and J. Woodburn (ed). Oxford University Press.

1991 Pragmatic Meanings of Possession in Paliyan Shamanism. *Anthropos I* 86 (367-84).

1994 Dimensions of subsistence foraging in South India *Ethnology 32.*

1994 Illness and Response Among South Indian Foragers *Medical Anthropology: Cross cultural studies in health and illness Vol. 16, No. 2.*

**Guha, B.S.**

1973 *'Indian Aborigines and Who They are'* in Publication Division (Revised Edition). The Initial People of India, Ministry of Information and Broadcasting, Govt. of India.

**Guha R.C.**

1995 'Introduction' in R.C. Guha (ed), *Social Ecology* Oxford University Press, New Delhi.

**Gurumurthy**

1983 Enquiry Report by the Commissioner regarding alleged existence of bonded labourers in Srivilliputhur taluk.

**Gurumurthi and Batcha**

1984 Second Enquiry Report by the Commissioner regarding alleged existence of bonded labourers in Srivilliputhur taluk.

**Hardesty Donald L**

1977 *Ecological Anthropology* John Wiley & Sons. New York – 1977.

**Hiremath S.R., Sadanand Kanwalli and Sharad Kulkarni.**

1996 *All about draft forest bill and forest lands.* Samaj Parivartan Samudaya, Dharwad, Centre for Tribal Conscientization, Pune, et al.

**Hutton J.H. (ed)**

1935 - *Census of India* 1931, 1,3. Simla ; Government of India Press.

**Indian Express**

1983 *Paliyars' lot better now*, The third January.

**Iyer, L.K.A. Krishna**

1909 *The Cochin Tribes and Castes*, 3 Volumes.

1937 *The Travancore Tribes and Castes*, 3 Vol. Government. Press, Trivandrum.

**Jung Bruce De**

1971 *The History of Kodai Kanal* Diocesan Press Madras.

**Kaplan D. and Manners**

1972 *Culture Theory* N.J. Engle Wood Cliffs Prentice Hall.

**Kolenda P.M.**

1996 "Religion, Caste and Family Structure: A comparative study of the Indian Joint family" in Milton Singer (ed). '*Structure and Change in Indian Society*'. Rawat Publications Jaipur & New Delhi.

**Kothari Ashish, Neena Singh, Saloni Suri (ed)**

1996 *People & Protected Areas: Towards Participatory Conservation in India*. Sage Publication, New Delhi.

**Kothari Ashish**

1997 *Understanding Biodiversity: Life, Sustainability and Equity*. Orient Longman, Delhi.

**Lee B. Richard, Irven Devore (ed)**

1968 *Man the Hunter*, Aldine Publishing Company, New York.

**Levision David and Melvin Ember**

1998 *Encyclopedia of Cultural Anthropology* Henry Holt and Company New York.

**Luiz, A.A.D.**

1962 '*Tribes of Kerala*' Bharatiya Adim Jati Sevak Sangha, New Delhi.

**Madura Manual**

1891 Extract from the Madura Manual 1891 in '*Ethnographic Notes on Scheduled Tribes*'. *Census of India* 1961 Vol-9 Madras Pt. 5-B.

**Mahalingam, S.**

1990 *Role of cooperative organization in developing tribal economy*. Mittal Publications, New Delhi.

**Mandelbaum Wilfred**

1971 *Modern India's Economy*, A Bell and Howell Company, Columbus.

**Mandelbaum D.G.**

1977 Anthropology and the Challenges of Development. Paper presented at the plennary session of 10th International Ethnological and Anthropological Congress, New Delhi.

**Mathur Hari Mohan**

1990 *The Human Dimension of Development Perspectives from Anthropology'* Concept Publishing Company, New Delhi.

**Mehta B.H.**

1968 Forestry and Tensions in Tribal Area in *Applied Anthropology in India* L.P. Vidyarthi (ed) Kitab Mahal, Allahabad.

**Milton K. (ed.)**

1993 *Environmentalism: The view from Anthropology*, ASA Monographs 32. Routledge, London.

**Mohanty B.K.**

1995 'A Paliyan folk tale and its Analysis: A discussion in the context of Emic Approach. *In Journal of the Institute of Asian Studies Vol.XII No. 2.*

**Morris Brain**

1978 '*Are there any individuals in India. Eastern Anthropologist,* 31: 365-77.

1977 'Tappers, Trappers and the Hill Pandarm (South India) *Anthropos* – 72.

1982 *Forest Traders: A Socio- economic study of the Hill Pandaram.* London Athlone Press.

**Mukherjee R.K.**

1942 *Social Ecology*, Long Man's Green & Company, London

1994 An Ecological Approach to Sociology in R.C. Guha (ed) Social Ecology, Oxford University Press, Delhi.

**Nambiar, P.K.**

1961 *Census of India* Vol. IX, Part VB (I) 'Ethnographic notes on Scheduled Tribes' Madras Government Press.

**Nayak P.K.**

1988 Tradition and Development in Tribal Orissa: A Case Study of Dongaria Kandha In *Tradition and Development* Sachchidananda etal (ed). Concept Publishing House, N. Delhi.

**Nelson J.H.**

1968 *The Madura Country* Published in Madras.

**Netting R.M.**

1996 Cultural Ecology in *Encyclopedia of Cultural Anthropology* David Levinson & Melvin Ember (ed). Henry Holt and Company New York.

**Patel M.L.**

1983 *Planning strategy for Tribal Development*, Tribal studies of India series T 111, Mittal Publication, New Delhi.

**Patel Sujata**

1996 'Ecology & Development' *Economic and Political Weekly*. Vol-XXXII No. – 38.

**Pati R.N., B. Jena (ed.)**

1989 *Tribal Development in India,* Ashish Publishing House, New Delhi.

**Paulraj S.**

1995 *'Grizzled Giant Squirrel Sanctuary – Management Plan*. Government of Tamil Nadu.

**Poffenberger Mark, Betsy McGean (ed.)**

1996 *Village Voices Forest Choices: Joint Forest Management in India*,Oxford University Press, New Delhi.

**Poffenberger Mark and Chhatrapati Singh**

1997 'Commonalties and the State: Re-establishing the Balance in Indian Forest Policy' in Mark Poffenberger et al (ed) *Village Voices, forest Choices Joint Forest Management in India*. Oxford University Press, Delhi. His quote from *W.R. Bentley* etal (ed.) 'Tenure and Agroforestry Potentials in India" is of special interest.

**Prasad U.L. and Rao U.M.**

1992 'Milch Animal Insurance " A Process Analysis' *Agricultural Situation in India,* January, Ministry of Agriculture, New Delhi.

**Publication Division**

1955 *The Tribal People of India*, Publication Division, Ministry of Information and Broadcasting, Government of India.

**Puri, G.S.**

1960 *Indian Forest Ecology*, 2 Vols. New Delhi: Oxford Book Company quoted by Gardner P.M. (1969) in Ecology and Social Structure in Refugee Populations: The Paliyans of South India.

**Rajkumar R. and Bal Sundaram Chellam**

1995 'Communicable Diseases Among Paliyan of Kodaikanal with reference to Health Care Access' *Journal of Indian Anthropological Society.* Vol-30, No.2.

**Rangarajan Mahesh**

1996 *Fencing the forest: Conservation and Ecological Change in India's Central Provinces 1860-1914,* Oxford University Press, New Delhi. His quote from (I) Rama Chandra Guha's An Early Environment Debate: The making of the 1878 forest Act,; I.E.S.H.R. Vol. XXVII (1990) PP, 65-84.

**Rath, S.N. and D.K. Behura**

1984 Man nature Relationship: Life in a Kisan Hamlet of Western Orissa. *Man and Life, 2 (3 and 4).*

**Roy Burman, B.K.**

1968 "Forest and Tribals in India in *Applied Anthropology in India*, L.P. Vidyarthi (ed) Kitab Mahal, Allahabad.

**Sahlins, Marshall D.**

1965 Culture and Environment: The Study of Cultural Ecology, *In Horizons of Anthopology.* Sol Tax etal (ed) George Allen and Unwin Ltd., London.

**Sahu Chturbhuj**

1996 *Birhor Tribe: Dimensions of Development*, Sarup & Sons, New

Delhi.

**Second Enquiry Report**

1984 Please refer to Gurumurthi & Batcha 1984.

**Sharma B.D.**

1984 *Planning for Tribal Development*, Prachi Prakashan, New Delhi.

1989 *The Web of Poverty*. Director of Publication N.E Hill Universty, Shilong.

**Shiva, V.**

1991 *Ecology and the Politics of Survival: Conflicts over Natural Resources in India,* Sage,New Delhi

**Shiva Vandana, Afsar H. Jafri, Gitanjali Bedi, Radha Holla-Bhar**

1998 *The Enclosure and Recovery of the Commons*, Research Foundation for Science, Technology and Ecology, New Delhi.

**Shore Cris, Susan Wright (ed)**

1997 *Anthropology of Policy: Critical Perspectives on Governance and Power*, Routledge, London.

**Singh Amar Kumar, M.K. Jabbi (ed)**

1997 *Tribal in India: Development, Deprivation and Discontent*, Har-Anand Publication, New Delhi.

**Singh S.P.**

1981 '*Tribal Development Administration*'. A historical overview. In L.P. Vidyarthi (ed) *Tribal Development and its administration*, Concept Publishing Company, New Delhi.

**Sol Tax**

1968 Action Anthropology in L.P. Vidyarthi (ed) *Applied Anthropology in India*, Kitab Mahal, Allahabad.

**Srivastava A.R.N.**

1990 Rise of Ecological Studies in Cultural Anthropology, *Man in India* Vol.-70 No. 3.

**Sukumar B. and Ahalya Sukumar**

1968 "A land sat Study for Eco-Development strategy around Palani

Hills for Western Ghats in Tamil Nadu", 6th Asian Conference on remote Sensing, ACRS, PP 556-58 Deforestation and ecological imbalance in Palani Hills of Tamil Nadu (Unpublished).

(unpublished) *Deforestation and Ecological Imbalance in Palani Hills of Tamil Nadu.* C.E.S.S. Thiruvanathapuram.

**Sundaram S.**

1986 *Voluntary Agencies and Rural Development* B.R. Publishing Corporation Dlehi.

**The Hindu**

1984 *Enabling them to lead a decent life*, the 12th October.

1987 *Bringing them into the mainstream*, Friday the 26th June.

**Thurston, E**

1909 *Caste and Tribes of Southern India Vol-V (1975 reprint)* Cosmo Publication Delhi

**Tracy J.E. (Quoted by Thurston)**

1909 Caste and Tribes of Southern India, Vol-V (1975 reprint) Delhi Cosmo Publication.

**Vidyarthi L.P.**

1963 *The Maler: A Study in Nature-Man-Spirit Complex of a Hill Tribe in Bihar*, Bookland Private Ltd., Calcutta.

1968 *Applied Anthropology in India ; Principles, Problems and Case Studies*, 'Kitab Mahal, Allahabad.

**Vidyarthi L.P.., Binay Kumar Rai**

1976 *The Tribal Culture of India*, Concept Publishing Company, New Delhi.

**Vidyarthi L.P. (ed)**

1981 *Tribal Development and its Administration*, Concept Publishing Company, New Delhi.

**Wyck off Charlotte Chandler**

1945 *Kodaikanal* 1845-1945, London Mission Press, Nagercoil.

# Index

**G**



**H**

**U**



**V**



**W**

# Photographs

***A Paliyan digging tubers with a digging stick (Shenbagathope)***

***Paliyan collecting clean water***

***Rock shelters used by Paliyans in earlier days (Shenbagathope)***

***Traditional housing of Paliyans (Palamalai)***

***A Dhevar medicine man married to a Paliyan (Mahalingam Kovil)***

***A Paliyan lady with Dhevar husband and children (Palamalai)***

***Paliyan deity being worshipped by Paliyans and non-Paliyans (Pilavakkal Dam)***

***A Christian festival being observed by Paliyans and Non-Paliyans (Palamalai)***

*An existing school at Shenbagathope*

*Kalyanmandap/ Community hall at Shenbagathope*

***Second phase houses at Shenbagathope***

***Dilapidated houses at Shenbagathope (Third phase)***

***Paliyans extracting barks (Pilavakkal Dam)***

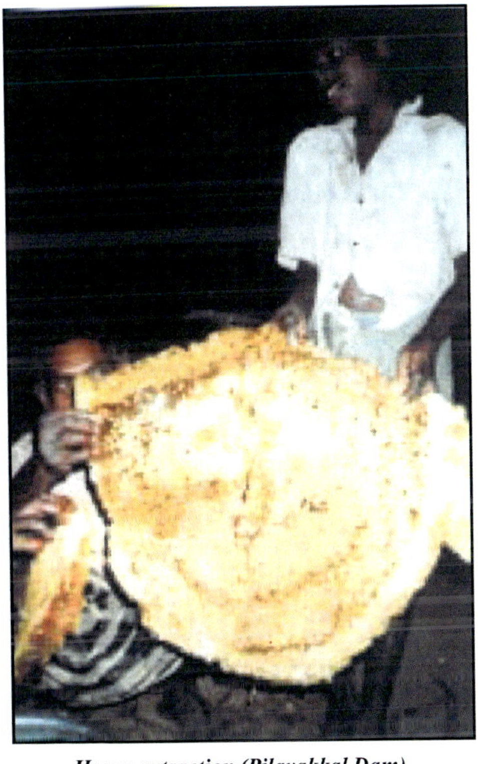

***Honey extraction (Pilavakkal Dam)***

***Water supply arrangement at Shenbagathope for Paliyans***

***A tea stall at Shenbagathope***

A Paliyan habitat (Mahalingam Kovil)

Paliyan children in a church at Palamalai

***Paliyan and non-Paliyans working together (Palamalai Estate)***

***An aerial view of Palamalai estate***

***A Paliyan habitat (Mahalingam Kovil)***

***Paliyan children in a church at Palamalai***

***Paliyan and non-Paliyans working together (Palamalai Estate)***

***An aerial view of Palamalai estate***